The American Mediterranean

Marilyn E. Weigold

KENNIKAT PRESS
Port Washington, N.Y./London
1974

The American Mediterranean

An Environmental, Economic & Social History of
LONG ISLAND SOUND

Library of Congress Catalog Card No: 74-80588
ISBN: 0-8046-9064-2

Manufactured in the United States of America

Published by
Kennikat Press Corp.
Port Washington, N.Y./London

TO THE MEMORY OF MY FATHER

The dangers of Whaling (Harper's Weekly)

Contents

Preface

THE AMERICAN MEDITERRANEAN GREW OUT OF A COURSE ON THE HISTORY AND ECO-
nomic development of Long Island Sound given by the College of White Plains in
cooperation with the Rye Historical Society in January 1973. The course consisted of a series of lectures on the Sound's past given by the author in the
fittingly historic setting of the Square House, the eighteenth-century Post Road
inn now used as the headquarters of the Rye Historical Society. Field trips to
historic restorations and business corporations in the Sound shore communities
of Connecticut, Long Island, and Westchester supplemented the lectures. Given
the proliferation of museums in the Sound region during the past decade, planning
the field trips was the least difficult aspect of the course coordinator's job. Far
more challenging was the problem of preparing a series of lectures. The search for
material pointed to the lack of a comprehensive history of Long Island Sound.

Thence emerged the present work, whose preparation was facilitated by the
generous assistance of the following people: Mr. Richard Winsche, historian of
Nassau County; Mrs. Mary Louise Matera of the Nassau County Museum Library;
Mrs. William Carpenter, Director of the Rye Historical Society; Dr. Lee F. Koppel-
man, Executive Director of the Nassau-Suffolk Regional Planning Board; Dr.
Edith Tannenbaum of the Nassau-Suffolk Regional Planning Board; Mr. David
Burack, Project Manager of the Long Island Sound Study of the New England
River Basins Commission; the Honorable Robert Moses; Commissioner Charles
E. Pound of the Westchester County Department of Parks, Recreation, and Con-
servation; Mrs. Doris Williams of the Environmental Information Service, State
University of New York at Stony Brook; Mr. and Mrs. F. Trubee Davison of Pea-
cock Point, Long Island; Mr. George Wilson of New Rochelle; Mr. William
Dornbusch of the Westchester County Historical Society; Mr. George O. Pratt of
the Fairfield County Historical Society; Mr. Stanley Crane of the Pequot Library;
Mrs. Agnes Packard of the Huntington Historical Society; Mr. Franz Edson of
Lloyd Harbor, Long Island; Mr. David Clemens of the Oyster Bay Historical
Society; Mrs. Beatrice Hrubek and Agnes Murphy of the Westchester Federated
Conservationists; Mrs. David L. Gillispie of the Oysterponds Historical Society,
Orient Point, Long Island. I would also like to thank the very capable staffs of the
Marine Historical Association, of the Roslyn Library, as well as those of numerous
government agencies listed in the bibliographic note.

The American Mediterranean

The Connecticut River wends its way towards the Sound (Photo by the author)

1

Discovery and Settlement

Europeans Come to the Sound

To boatmen and sailors who know Long Island Sound, it might not be surprising to learn that Daniel Webster called the Sound the "American Mediterranean." Perhaps the first European to sail on this waterway, the Dutch captain Adriaen Block, might have had similar impressions in 1614 when he traveled from Hell Gate eastward through Long Island Sound. This was not his first visit to these shores. He had crossed the Atlantic before and sailed along the east coast from Maine to Delaware, seeking contacts with the natives. Coming here again in 1613, he was able to establish friendly relations with the Indians of Manhattan Island, trading with them for furs. When he was ready to sail back to the Netherlands, fate interfered with his design. Fire of mysterious origin destroyed his boat, furs and all. Seemingly a cruel blow, Block's misfortune actually turned out to be a blessing in disguise.

To be able to get back home after the fire, the resourceful captain and his crew had to get busy and build a new ship. They aptly called it the *Restless,* possibly the first vessel constructed by Europeans in America. Was the new boat weighing only sixteen tons seaworthy enough to make the Atlantic crossing? To answer this question, its builders decided to test their own handiwork by sailing the waters around Manhattan. Venturing up the East River in the spring of 1614, the frail craft suddenly found itself in a torrent of swirling currents to be called appropriately, Hell Gate. Some years later Daniel Denton, an English visitor, described the fury of the spot in this vivid language:[1]

There runneth a violent stream both upon flood and ebb, and in the middle lieth some islands. Rocks, which the current sets so violently upon, that it threatens present ship-wreck; and upon the flood is a large whirlpool which continually sends forth a hideous roaring, enough to affright any stranger from passing any further . . .

The *Restless* went through this aquatic baptism of fire no doubt as much with good luck as with the courage and ability of skipper and crew. At any rate, fate, so cruel when the flames consumed Block's first ship and cargo, now lent a helping hand. Passing through Hell Gate without a mishap, the brave Dutch captain sailed into fame by discovering Long Island Sound in all its unspoiled and unpolluted beauty, and what a sight it must have been then!

Sailing eastward on the Sound, he passed the wooded shores of Westchester County and Connecticut, darting in and out among the Norwalk islands and coming upon the picturesque rivers which flow into the Sound. The broad estuary of the Connecticut was too tempting not to be explored, and the *Restless* sailed up the river beyond the site of present-day Hartford. Blocked by shallow waters, the ex-plorers turned around and, rounding the end of Long Island, landed at Montauk Point. From there they continued their voyage past Block Island, named for the commander of the *Restless.* The ship then went to Cape Cod where Block boarded another Dutch vessel for the trip back to the Netherlands, leaving command of the *Restless* to Captain Cornelius Hendricksen, his subordinate. Hendricksen is believed to have made additional explorations on the Long Island side of the Sound before returning home in 1616. In the meantime Block had appeared before the States General of the Netherlands, describing his explorations with the help of a map which traced the shoreline of the Sound fairly well, indicating the complete separa-tion of the "long island" from the continent. The "American Mediterranean" of the New World had not only been discovered but also properly mapped.

Several years later Thomas Dermer, an Englishman, sailed through the Sound en route to Virginia. He too had to have the skill, determination, and luck of the early explorers, because navigating the Sound in the days before charts was risky business. Netherlanders and Englishmen who would subsequently make their homes either on the Sound's mainland shore or on the North Shore of Long Island would quickly realize this. But they were not deterred, because the Sound's advantages far outweighed its terrors. One of the great benefits of living near the Sound in the seventeenth century was the limitless supply of food offered by its clear waters. On this subject the Dutch explorer Adriaen van der Donck commented:[2]

Those fishes are so tame that many are caught with the hand . . . In some of the bays of the East River the cod fish are very plenty; and if we would practice our art and experi-ence in fishing, we could take loads of cod fish, for it can be easily accomplished.

Another advantage of living near the Sound was the fact that the waterway served as an avenue of transportation at a time when overland travel was virtually

4

impossible. A seventeenth-century Dutch manuscript sums up this benefit quite accurately by stating:[3]

. . . from the river Mauritius [Hudson] off to beyond the Fresh River [Connecticut] stretches a Canal that forms an island forty miles long, called Long Island, which is the ordinary passage from New York to Virginia having . . . many harbors to anchor in so that people make no difficulty about navigating it in winter.

For settlers both on the mainland and on the "long island" which lay across the bright blue water, the Sound was the connecting link and on its shores there developed a rather distinctive civilization which bore the marks of the sea. Its beginnings go back to the time when New Amsterdam, the predecessor of New York City, was a Dutch colony on the tip of Manhattan Island. Yet New Amsterdam had little or no effect upon the distant English-speaking settlements to the east, whose hardy pioneers had to eke out a living on the shores of the Sound, relying entirely upon themselves.

The Coming of the English Settlers

The first comers were twenty men selected in 1635 to go from Hartford to the mouth of the Connecticut River where the town of Saybrook lies today. The tasks of the little group led by Lion Gardiner were to take possession of an abandoned Dutch fort, build permanent shelters for themselves, and then fortify the place. After its completion the fort was to play an important role in the war against the neighboring Pequot Indians waged in 1637 by the settlers of the Connecticut valley. This warlike tribe, inhabiting the area east of the Thames River, had spread terror far beyond their home territories intimidating other Indian tribes as well. But under the leadership of John Mason and Captain John Underhill the whites practically destroyed them. After this danger was eliminated, the shoreline of the Sound became free for settlement.

While the Pequot War was still in progress, two ships bearing a group of religious dissenters known as Puritans arrived in Boston. Headed by the Reverend John Davenport, the group included the prominent merchant Theophilus Eaton. Hoping to find a new home among fellow Puritans in Massachusetts Bay, the newcomers soon discovered that considerable dissension existed among the inhabitants of the colony. Seeking to prevent their own group from being torn apart by these quarrels, Davenport and Eaton began to search for a new home for their people. They had heard glowing reports about the lovely area on the Sound near the mouth of the Quinnipiac River. The merchants of the group, already contemplating lucrative trade with London, were especially interested by the news that this area had a deep-water harbor. The attraction of this Sound-side paradise was

5

so irresistible that the group left Boston at the end of March 1638 and arrived at their destination on April 24th. Immediately the hard work of building temporary shelters began. Pits and caves were dug, the walls fortified with wood, and the elements kept out by a roof covered with thatch. They called their community New Haven, a name which was more a projection springing from unshakable faith than any reference to fact.

In addition to providing shelter for themselves, the settlers also swore a solemn covenant to follow the Scriptures in all their future work while erecting an ecclesiastical and civil government. All this would result in a well-ordered society, thought by the merchants of the group to be necessary for economic survival. An even more important factor of survival was military defense. Every man capable of bearing arms was ordered to equip himself with a musket and ammunition and participate in weekly military exercises. The seventeenth-century form of conscription did not apparently deter many people from settling in New Haven, because five years after its founding the colony had eight hundred inhabitants.

During the same five-year period several other settlements sprang up under the sponsorship of New Haven. All of them but Southhold, founded across the Sound on Long Island, were in Connecticut, that is, Guilford, Branford, Stratford, and Fairfield. Robert Feaks and Daniel Patrick, exploring the area along the coast, purchased land from the Indians and founded Greenwich. The Captain Islands in the Sound off Greenwich were named for Captain Patrick. This doughty explorer, subsequent to the purchase of land, had much trouble with the Indians. In 1643 he led an expedition against them, but to the dismay of the Dutch and English participants no Indians could be found. When the troops were disbanding at Stamford, one of them, a Netherlander, accused Captain Patrick of misleading the expedition and killed him in the ensuing heated altercation. The Dutch eventually had better luck than Captain Patrick in their attempt to suppress the Indians. An expedition across the Sound to Cow Bay (Manhasset) was successful. A combined English-Dutch militia under the command of Captain John Underhill, who had relocated in Stamford following the Pequot War, defeated the Indians near Maspeth.

In eastern Connecticut the severe Indian menace slowed the spread of white settlements. The town of Mystic, Connecticut, could not be founded until 1654. In that same year war broke out between the Pequots and Narragansets of New England and the Montauk Indians of Long Island. Attempting to bring the Long Island Indians under their domination, the fierce Pequots staged raids across the Sound. One of the excursions yielded quite a prize, as the raiders seized the daughter of chief Wyandanch of the Montauks on her wedding night. The unfortunate bridegroom lost his life in the struggle, but Lion Gardiner, due to his former connection with the Narragansets, was able to intercede and eventually restore the bride to her parents. This friendly intercession, plus the fear of the more powerful

6

tribes across the Sound, forced the Long Island Indians to remain on fairly friendly terms with the white man. The English settlers, however, did not always trust them. During King Philip's War of 1675, for example, the English governor of New York ordered that all canoes east of Hell Gate "be seized and delivered to constables" while "all such canoes as should be found in the Sound after that time should be destroyed."[4] The governor feared lest the Long Island Indians cross the Sound to help their allies on the mainland. This could be easily done in view of the fact that some of the tribes inhabiting the Sound shore area had canoes capable of holding up to forty people. With such equipment war parties could have been easily transported from one side of the waterway to the other. But in reality the governor did not have much to worry about, because the Long Island Indians remained quiescent.

It was only natural that the inhabitants of Connecticut would become interested in the island across the Sound. In 1633 Governor Winthrop of Connecticut sent a ship, the *Blessings of the Bay* out to explore it. The modern Argonauts returned with vivid descriptions of what they saw. Another enterprising man from Hartford, Sam Wyllys, acquired Plum Island off Orient Point in 1659 from Wyandanch, sachem of the Montauk Indians. Unlike the purchase of another island, Manhattan, this one involved no cash payment, just "one coat, one barrel of 'biskitt' and one hundred . . . fishhooks."[5]

Other mainlanders setting out across the Sound from New England acquired Setauket and Eaton's Neck on Long Island. During the colonial period Setauket, Flushing, Matinecock in the vicinity of Oyster Bay, Huntington, Smithtown, and Wading River were the scene of Quaker meetings. The Quakers established settlements on the mainland as well as at Mamaroneck on the Sound and farther inland in Westchester County at Purchase and Chappaqua. Visitations to the far-flung meetings on both sides of the Sound were made by such Quakers as Elias Hicks, who passed back and forth from the mainland to the island on numerous occasions.

As for Eaton's Neck, Theophilus Eaton, one of the founders of another religious settlement, Puritan New Haven, purchased this area, originally an island off the North Shore of Long Island, in 1646. Farther east on the island Richard "Bull" Smith obtained Smithtown in a rather curious way. Legend says that he made a little deal with the Indians whereby he would acquire as much land as he could cover in one day riding a bull. The story has been commemorated by a gilded statue representing the crafty negotiator riding on his charger. The statue stands in a park overlooking the highway leading from the west to the little Long Island metropolis to which the intrepid rider gave his name. These early English settlers, like those who founded Port Jefferson after purchasing land from the Indians, imparted a definite character to the North Shore and north fork of Long Island. To this day the homes, churches, and the general layout of the villages resemble New England towns of the seventeenth and eighteenth centuries.

7

The western portions of the island had been subjected to Dutch as well as to English influence, and not until 1650 did a firm dividing line drawn at Oyster Bay separate the two colonies. In 1664 when New Amsterdam surrendered to the English, all of Long Island, including the towns on the eastern end which were geographically and economically linked to Connecticut, became part of the New York Colony. Eight years later the Dutch recaptured New York and held it for fourteen months before returning it to England under the terms of a peace treaty. In the interim Long Islanders had to contribute money for the defense of their towns against the Dutch, but most refused, demanding a representative assembly as the price of their contribution to the defense fund.

Smuggling and Pirates

In other ways, too, Long Islanders were thwarting the British. Instead of bringing imported goods into New York harbor, they were smuggling them into Sound harbors on the North Shore in order to avoid paying duty. When the Earl of Bellomont, governor of the New York Colony, was charged with stamping out smuggling, he concluded that Long Islanders were lawless and unruly and that the area was a receptacle for pirates. Ironically, in an attempt to suppress piracy the earl appointed Captain William Kidd to command the ship *Adventure Galley* whose mission was to stop illegal activities. The appointment became the great temptation. The once respectable Kidd who had lived in Manhattan with his wife now engaged in various forms of piracy on the world's seas. The smuggling done by Long Islanders, although far from commendable, could not be compared to the type of lawbreaking attributed to the infamous Captain Kidd. He became the terror of the Sound, preying upon commerce and stopping here and there to seize supplies.

Pirates' exploits always appeal to popular imagination, giving rise to romantic stories. Hidden treasures are a favorite theme. The greater the buccaneer's fame, the more persistent the legend about the mysterious hoard. Captain Kidd is said to have buried a treasure chest on Gardiner's Island, warning the earl that either he or his son would literally lose his head if the chest was not there when he returned for it. It was an empty threat. The captain, expecting to obtain a pardon, surrendered to the Earl of Bellomont. Sent to England for trial, he was hanged in 1701.

As for the rest of Captain Kidd's treasure, weekend explorers are still looking for it everywhere from Kidd's Rock in Manhasset to Sachem Head's Harbor, Guilford, the place so named because an Indian sachem was killed there as he tried to swim away from his pursuers during the Pequot War. In the Thimble Islands off Branford, Connecticut, there is a pirate's cavern, and in one of the rock formations a carved-out area called Captain Kidd's Punch Bowl, but thus far no treasure has been uncovered.

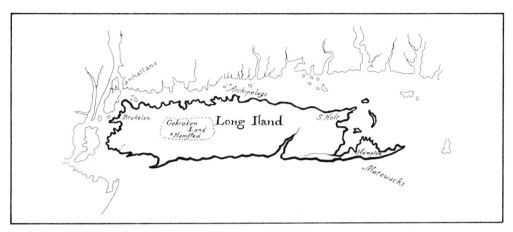

Detail of Map by Nicollaum Visscher of Amsterdam 1662 ("Journal of Long Island History," 1967)

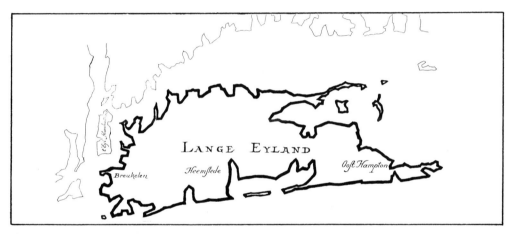

Detail of Map of "Niew Nederland" by Johannis Van Kevlin of Amsterdam, 1685 ("Journal of Long Island History," 1967)

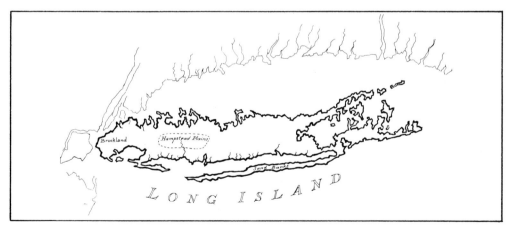

Detail of Map published in London, circa 1720, an accurate copy of the Morden Map ("Journal of Long Island History," 1967)

Seining for menhaden (Harper's Weekly)

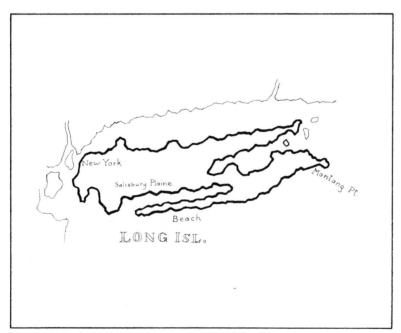

Detail of Map by Henry Moll of Augsberg, 1729 ("Journal of Long Island History," 1967)

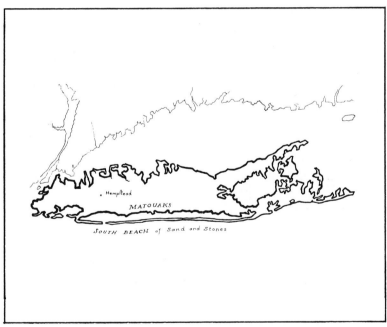

Detail of "A Map of the Most Inhabitated Part of New England" by Thomas Jeffreys of London, 1755 ("Journal of Long Island History," 1967)

Cold Spring Harbor Whaling Museum (Photo by Allan M. Eddy Jr.)

Detail from diorama at Whaling Museum showing steamship towing passenger sailing vessel (Photo by Allan M. Eddy Jr.)

Probably the only good thing which can be said about Captain Kidd is that his death ended an era on the Sound. It had the happy effect of deterring others from imitating him, with the result that as the eighteenth century dawned the Sound was perhaps safer than ever before. Not only was piracy on the decline, but the inhabitants of the Sound shore communities had come to know their "American Mediterranean." No longer was the Sound an uncharted sea, nor for that matter was the land on both sides of the waterway unexplored.

The Spread of Settlements

By the early eighteenth century settlements had been established along the North Shore of Long Island, the Connecticut coast, and farther down in Westchester County all the way to the Sound's eastern gate at Throgs Neck. The natural attractions of this area were described by the Dutch as follows:[6]

The country on the East River between Greenwich and the island Manhattans, is for the most part covered with trees, but yet flat and suitable land, with numerous hills and valleys, right good soil for grain, together with fresh hay and meadow land.

In 1640 the Dutch bought land in this area from the Indians and called it Vredeland, or land of peace, a misnomer if there ever was one. Perhaps the Dutch were impressed by the peaceful aspect of the scenery, the gently rolling hills covered with woods or lush grass, and the meadows extending from what is called today the Hutchinson River to the seashore. To this land came religious refugees, among them the famous dissenter Anne Hutchinson and the Baptist John Throgmorton. Both came with followers who had joined them in Rhode Island, another place of refuge for people who were dissatisfied with conditions in the Massachusetts Bay Colony. It has been said that Anne Hutchinson, when rumor spread in Rhode Island that she would be brought back to Massachusetts and tried for witchcraft, decided to move farther and seek safety from persecution under Dutch authority.

The two parties, Hutchinson's and Throgmorton's, arrived in their new locale in 1643 after they had received patents from the Dutch. The Indians in the neighborhood, still resentful of what they thought was gross mistreatment on the part of the Dutch, broke the peace and attacked the English settlers, killing most of them, including Anne Hutchinson. Throgmorton fled on a boat that just happened to sail by when the Indians staged their raid. No remains of the settlement could be found, and only a few names commemorate these early pioneers. Throgmorton gave his name to Throgs Neck and later to the Throgs Neck Bridge, while the little brook serving as the border of the settlement has been known as the Hutchinson River; later the Hutchinson River Parkway became the ultramodern artery of leisurely motor travel in the late twenties—two tributes of the automobile age to the age of the intrepid pioneers.

Another early settler, Thomas Pell, came from Fairfield, Connecticut, and bought land from the Indians in 1654. He had better luck than his predecessors in the area. Building a home near the Sound, he practically invited the Dutch from New Amsterdam to attack him, but he was left alone. Nor did the Indians molest him. Pell and his heirs simply went about their business conveying pieces of land, including City Island and New Rochelle, but not all the Pells fared well. One of the heirs, John Pell, drowned when his boat went down off City Island in a fierce wind in 1702, but somehow the Pell family managed to survive. In this sense, they were not terribly different from the Huguenots to whom they sold New Rochelle. Having come to the New World to escape religious persecution in France, the Huguenots, particularly those who emigrated from the French coastal town of La Rochelle, may have been attracted by the geographical position of their new home town on the Sound.

The Sound may also have been a factor in the settlement of Mamaroneck. John Richbell, who bought Mamaroneck from the Indians in 1657, had originally come from Barbados and had more recently lived in another coastal area, this one on the Sound, Oyster Bay. He was not the most popular man on either side of the Sound, for he quarreled with his tenants and neighbors east and west over boundaries and caused resentment, especially among people in Rye.

The origin of this town goes back to 1660 when a group of men from Greenwich purchased Manursing Island from the Indians. They chose this island, located in the Sound but very close to the mainland, because of its distance from both the Dutch in New Amsterdam and the British in the Connecticut Colony, and also because of its relative proximity to another Sound shore community, Greenwich. Still feeling close to their old home, the men who purchased Manursing Island felt that in the event of difficulty in Rye they could, by sailing only a short distance, retreat to Greenwich. Within a short time the original settlement spread from the island to the mainland, and while there was a long struggle to determine whether Rye would be part of New York or Connecticut, the town grew and prospered. That conflict was finally resolved in 1700 in favor of New York.

The Pioneering Era

Quite aside from boundary disputes and occasional trouble with the Indians, the young settlements on the Sound faced very serious problems during the early period when just staying alive meant a never-ending struggle. Even after the newcomers had built permanent homes for their families, life remained harsh and demanding. Almost continuously these hardy pioneers had to cope with the problem of how to stave off hunger. If they arrived in the spring and planted crops, the harvest was many months away. In the meantime food had to be obtained. Residents living near the sea had it easier. They could always find fish, clams, and oysters in abundance. On both sides of Long Island Sound, the many inlets and sheltered coves with their shallow waters offered them all manner of marine life. The rivers flowing into these inlets had other varieties of fish. If tired of sea food, the fisherman, turning into a hunter and going into the nearby woods and fields, could bring home deer, rabbits, or other game, though this way of providing food was more time-consuming and strenuous than dipping into the sea.

Despite the variety and abundance of fish and game, the settler had to rely on agriculture as the mainstay of his life. Often, to assure open land for cultivation, the forest had to be cleared; this meant back-breaking work that had to be done mainly with the help of primitive tools and draft animals if they were available. After wielding the ax for hours, after clearing away the branches and the logs, settlers had to face the heavy, immovable waist-high stumps anchored deeply in the rocky ground. Without special tools or dynamite, the stumps, like silent mementos of the forest, had to be left standing. By plowing around them, the pioneers could cultivate patches of land, and the fertile soil of the forest, its dense canopy removed to let in the sun, would compensate so much hard labor with a rich harvest.

Another phase of the early settlers' work was to care for their domestic animals, the principal source of food for the community. But some of them did not require much care. Pigs, for instance, the most important livestock in the economy of the early settlements, could forage in the forest on roots and nuts and could take care of themselves if attacked by wolves or wildcats, the most common predators of the region. Sheep, however, would fall easy victim and, along with the cows, had to be kept near the dwellings. To guard their possessions, many of the communities employed herders on the common grazing grounds. There were no cowboys riding horses. The horse was a luxury in this period when people traveled mostly by boat. On the farms oxen rather than horses were used as draft animals. Work was heavy and incessant, and the whole family had to participate, including children. Indeed, the prevailing attitude among the early settlers was "the more, the merrier" when it came to having children. Even small tots could be trained to perform chores while the older children could work side by side with their parents, helping with the general operations of the farm.

The Development of Trade

Trade is the lifeblood of civilization. It breaks down the barriers of isolation, stimulates new ideas, and by exchanging the surplus products of different areas helps to raise the standard of living. All these influences were now slowly percolating through the tiny shore communities along the Sound, where the life of the early

11

pioneers was so hard, drab, and tiring. Occasionally, however, the monotony might be broken by the arrival of a traveler from another Sound shore community, though in the early days this was uncommon. Actually, little contact of either a social or an economic nature existed among the different settlements. Trade during these early pioneer days was still in its infancy. Furs were an exception, for which the Indian obtained a few pieces of clothing, tools, or just trinkets. Inland communication was difficult unless the hunter or trapper, blazing the way to the future forest tracts, would follow deer paths or Indian trails. Rivers were the only way to penetrate deeper into the interior, but all rivers ran contrary to the lateral trade that might connect one settlement with the other. Only the sea was available for such contacts. Each village, therefore, constructed a dock or landing to accommodate any boat which might bring a trader offering perhaps just a scanty assortment of goods, besides the news and rumors always welcomed by the isolated settlers. They themselves produced only enough to satisfy their own needs and had little left over to exchange in trade. Cattle were the most important commodity; in order to encourage the cattle trade, New Haven, the largest town on the Sound, introduced annual fairs or markets as early as 1644. Here whatever surplus existed in the region could change hands. Money being scarce, people carried on trade by barter or used Indian wampum, a decorative string or belt made of colored seashells, as a means of exchange. Merchants engaged in this earliest trade would buy up the beaver skins, the corn, biscuits, bacon, or cattle and take them to New Amsterdam, to other areas on the Sound, or, as was usually the case, to the Bay area of Massachusetts.

Tools, fineries, liquors, arms, powder and lead, and boats were the most frequently traded goods. In the case of New Amsterdam this activity grew important enough by 1656 for the town to appoint a broker, John Pecq, who spoke both Dutch and English. To encourage this trade, the director of the Dutch colony, Peter Stuyvesant, opened annual fairs for cattle and sent out English-language placards to the Sound area to be posted by the local authorities. But the high duties on imports as well as exports collected in New Amsterdam stood in the way of a brisker development of trade. Not until the English permanently occupied and administered the Dutch colony after 1664 did New York become the great magnet drawing the trade of the Sound increasingly to its wharves, fairs, and counting houses. Until then the Bay area of Massachusetts remained the center of trade, especially for the eastern part of the Sound.

The dominance of Boston in this wider exchange of goods irked the merchants of New Haven, who would have liked to make their town another center of this overseas trade. They were unable, however, to overcome certain initial difficulties, among them the loss at sea in 1646 of the ship intended to initiate this trade. Yet New Haven could have initiated a lucrative trade, and not just with other colonies along the Sound. Settlers living as far away as Virginia would have been potential

12

customers of the New Haven merchants. Indeed, Virginia was becoming rich on her tobacco crop and had money to spend. Her inhabitants needed all kinds of goods for which they could pay cash. Once past the initial phase of colonization, some of the Sound shore communities could have supplied Virginia with the goods she needed and at a lower price than the mother country, especially with avoidance of duty, a practice widely indulged in by the colonists.

Breakdown of the Old Order

By the time the Sound shore communities had developed to the point where they had surplus goods to trade, they had reached a new plateau of development. The pioneering stage with all its fundamental problems receded into the past. With the beginning of the eighteenth century, a new day arrived. The population by this time had grown considerably, so that some division of labor appeared and the practice of every man being a jack of all trades, a situation absolutely necessary at the start of colonization, could be abandoned. The sons and grandsons of the first settlers could diversify and acquire special skills and, if they wished to, could abandon farming altogether. They could also step out of the tight religious and civil organization created by the early British colonists.

The settlers arriving in the Sound area from the Bay Colony of Massachusetts or directly from England, as we have seen in the case of New Haven, brought with them and continued to live by a closely knit religious organization. In it the civil and religious authorities acted as twins if not as hands in glove. Such close cooperation guided the destinies of the New Haven group, creating a "Bible state" where religion occupied the highest level in the hierarchy of values. The principle of the rule of law, so deep seated even today in the American political ethic, strictly defined the limits of human action. According to the Puritans, strict observance of the law had to be enforced in order to prevent the devil's work from undermining the social order. Uncontrollable human passions, unless kept within limits by those whom divine providence designated to be the leaders of society, could weaken, even destroy, the orderly functioning of their communities. Guided by moral values and rigorous standards observed through a strict code of behavior, they aimed at setting up a well-regulated homogeneous society of the faithful, a true "new haven," shielded by God's rule of law from the storms of human passions and dissensions.

When the original group of Puritans arrived in any of the Sound shore communities, the first thing they did, after taking care of the most urgent tasks, was to build a meeting house where services would be held and the affairs of the community discussed. The meeting house occupied the central position in the community. The houses of the inhabitants were built around it, and the plots of each family extended outward into the fields. As living conditions improved and the families increased in size and wealth, more lands were allotted out of the common property purchased from the Indians. Either the new lands thus received by each family were added to the original holdings, increasing the length of the strip, or the recipient might decide to take his allotment in the form of a parcel somewhere else, perhaps at a distance from his original lot. This in time resulted either in long strips extending far out into the rocky uplands or forests, or in a mosaic of various pieces scattered all around. Purchases, inheritances, marriages added to the irregularity of the pattern. Farming such widely separated parcels or extremely long strips was difficult at best.

The roads usually radiated from the central square, extending into the countryside. Adjacent to the meeting house real estate values rose. Here were the allotments originally distributed at the time of the settlement. As population increased, holders of the central area grew wealthy, could hire labor, or would if opportunity arose buy up the allotments of those whose land lay not too distant from the center. Thus, central holdings could be rounded out, increasing the wealth of the owner even more. The seller, in turn, could buy much more acreage farther out at bargain prices. These land deals would spread out the community so far that the original, tightly-kept organization would loosen up, ending in final separation. Since great distances from the center caused problems, there were many arguments and open quarrels in the meeting houses. Material considerations began to undermine the Puritan spirit which had guided the original settlers to found a new Zion. Indeed, the trouble with these first Bible societies was that they, as a community, assumed too much responsibility in guiding the lives of the members. Then, with increasing wealth, the desire for a more comfortable life was too great. In short, the descendants of the settlers had a different outlook in the eighteenth century. Economic opportunities were too tempting. With trade possible now in many places, the acquisition of better furniture, finer clothing, silverware, better tools, etc., acted as a powerful incentive to produce more, so that more could be offered for trade or sale. The community as a religious unit could no longer exercise any hold on its members who were lured by more land, more forest, and more meadow to produce more lumber and raise more livestock.

Eventually distances between the newly opened clearings and the center of town grew so much that physically all contact ceased and there was no way to enforce the regulations implemented by the Puritan leadership. The spiritual ties were severed, too. Swarming out to outlying lands, the settlers often became involved in disagreements over taxes, the opening of new roads, the erecting of new schools, and the hiring of new ministers. To solve their problems, they created new communities in which, without copying the tightness and closeness of social organization established by the original settlers, they preferred to have their holdings in one single area. An independent spirit developed among these people taking care of their holdings alone, in contrast to that of members of a tightly organized cooperative group directed by leaders claiming to derive their authority from God. The new farmer may have remained a good Christian, but he became an individual. The Yankee spirit began to emerge.

The Spread into the Interior

By the beginning of the eighteenth century this movement toward separation —that is, the establishment of new communities independent of the original settlements—became a common phenomenon on both sides of the Sound. But such changes took place amid heated arguments and factionalism often lasting for years. The old settlers put up a strong and bitter resistance to the separatist movement, because it meant the loss of taxes and other contributions to the community; but generally speaking, the fighting ended in the final breakaway of the "outlivers." New communities came into being like North Stamford, Bedford, New Canaan, Darien, Wilton, Redding, Ridgefield, Newton, and others. The settlers came to these places from the old communities on the Sound, mainly from Stamford, Norwalk, and Stratford.

Inevitably these migrations were responsible for the appearance of the first overland routes, initially just enlargements of old Indian trails. The interior of the country was opened up, however, and in time the old seashore settlements found themselves surrounded by a trading area. The farmers from the hinterland would bring their produce to the seaport town and make their purchases there. The volume of trade would increase as the roads penetrated deeper and deeper into the interior.

The Coming of Overseas Trade

The most lively example was New London, founded in 1646, whose citizens right from the beginning, took to the sea and became intrepid sailors. Boat building began here in 1756 by the launching of twenty- to thirty-ton vessels which could ply the coastal routes and later venture into the open sea as far as Nova Scotia and Newfoundland in the north and the Caribbean and the West Indies in the south. In the north they supplied the great fishing fleets with provisions, in the south the tropical plantations with dried or salt fish and other necessities. The rich farming area and the forests surrounding New London provided the cargoes making up this lively export trade. The principal items were barrel staves, smoked meat, corn, flour, and livestock. Horses were especially welcomed by the West Indian islands, and some boats would sail with a load of fifty to sixty animals on one trip. The produce would arrive in the harbor of New London in wagons drawn by four, sometimes six, horses or, if the hauling distance was not too great, by a team of oxen. The roads were usually in poor condition and travel on them always hazardous. The drivers and drovers would frequent the taverns around the harbor and often celebrate payday with drunken excesses, causing consternation among the peaceful citizens of the town. New London seems to have been an early forerunner of Dodge City or Abilene of the nineteenth-century cattle drives.

Since New London, lying near the eastern end of the Sound, enjoyed geographical advantages, the English government made it the only authorized port of entry at that time. Nevertheless, the trade of the town remained local in the sense that no transatlantic sailings originated here. Boston was the center of the transatlantic trade, where the imports from England arrived. Later New York became an ever-growing competitor, but the port cities of the Sound could not rival either of these great emporia of the North Atlantic.

New Haven was especially anxious to become a center, even if only a local one, in transatlantic trade, to break the monopoly of Boston and New York and share in the profits of the middlemen. In order to compete with her rivals, the town's assembly passed an act in 1747 to encourage direct trade with England by paying a bounty of five pounds for every hundred pounds of goods imported from and exported to England, but the measure failed. New Haven just could not generate enough trade to support the venture. Yet the town enjoyed rather prosperous conditions. The population, numbering fourteen hundred in mid-century, grew to eight thousand by the beginning of the Revolution, while the amount of tonnage in the same period increased fortyfold. In 1755 the first post office opened in town, and in the same year the *Connecticut Gazette,* a four-page weekly, began publication. Advertisements in it offered English and West Indian goods for sale for cash or for country produce. Among the goods offered were tea, brown sugar, rice, coffee, and spices. Finally, in 1756, the British government established a customs house in New Haven with jurisdiction extending from Killingworth to Greenwich. The port had over a hundred vessels, of which many were owned by fishermen. Ten per cent of the population—that is, 756 persons—were seafaring men.

In 1717 a stage service was granted to Captain John Munson to run coaches from New Haven to Hartford. The service could not operate in the winter months, and at any time the drivers had the right to order passengers off the stage if the going was too difficult, a good indication of the condition of the roads. Mail coaches started earlier. In 1684 the British crown granted the right to the colonies to run such a service between New York and Boston. The route passed through New Haven, New London, Stonington, and Providence as did later the New York, New Haven and Hartford Railroad in its heyday of development. The path followed by the royal mail service and later by the railroad and, not to be left unmentioned, by the New England Thruway, was originally blazed by the Pequot Indians, the very

16

first inhabitants of the region. Their path, like all Indian trails, followed the way of least resistance, winding around the hills and seeking fordable passages across the rivers.

The best way to travel in those pre-revolutionary days was on horseback, or on foot if the distance was not too great. Taverns were few and far between, and even though they were strictly regulated, they could offer only rather primitive accommodations. In fact, it was not until the turnpikes appeared after the Revolution that conditions for the traveler improved to the point where overland journeys ceased to be intolerable hardships.

To the inhabitants of the shoreline of the Sound, however, the condition of the roads mattered little because they traveled by sea if travel was necessary. For seamen engaged in trade the Sound offered a sheltered area protected by Long Island from the storms of the Atlantic. Here the smaller vessels, and there were many, could navigate with comparative safety. It was a different matter beyond Watch Hill and Block Island, where the wide open sea faced the sailor with all its hazards. For him, it was the school of hard tests which he faced with fortitude, because for him the sea and its terrors were not a deterrent but a lure. Whether fishermen, sailors of clipper ships, or whalers, they sailed forth from their "Mediterranean" to accomplish great feats, making their home towns on Long Island Sound famous round the world. But before they could sail out of their sheltered waters to the high seas they had to bear the burden of two wars with the mother country. Since for England command of the sea was of paramount importance in both wars, Long Island Sound bore more than an equal share of the suffering that befell the North American seacoast.

Early agriculture on Long Island (Huntington Historical Society)

2

Revolution and War
1775-1814

LONG BEFORE THE OUTBREAK OF THE REVOLUTIONARY WAR IN 1775, DISSATISFAC-
tion was rife in the thirteen British colonies of America. The inhabitants resented
the treatment they received at the hands of the despotic divine-right monarch King
George III. Restrictions of all kinds going back well over a hundred years, instead
of diminishing grew steadily, ending in imperial taxation considered by the colonists
to be oppressive and discriminatory. Inevitably such conditions found expression in
political unrest and agitation. Some demanded a voice in the government, while
others, taking matters in their own hands, went ahead and defied the laws formulated
and imposed upon them without their consent. Aside from taxation, trade restrictions openly favoring the mother country at the expense of the colonies, were
especially irksome. For instance, any ship sailing with the ultimate destination of
the North Shore of Long Island had first to put into New York harbor to pay duty
collected by the king's agent to enrich the royal treasury. Many inhabitants around
the Sound, in order to avoid paying heavy duties for carrying on legitimate trade,
simply took to smuggling.

The Sound Becomes a Theater of War

Of course, there were loyal British subjects who frowned upon such activity, and
when the Revolutionary War broke out many of them, known as Loyalists, or Tories,
cast their lot with Great Britain. This caused serious rifts in the Sound shore communities with the result that those who favored independence, the patriots, were at
odds with the Loyalists, including heretofore friendly neighbors and even family
members.

19

The break-up of family and community life reached the crisis stage when actual fighting came to the New York area in August 1776. The Battle of Long Island delivered a serious blow to the Americans, and if it had not been for George Washington's good fortune in finding dedicated men like Colonel John Glover with his Marblehead fishermen who ferried his troops across the East River in the dead of night, independence might have been lost. In his account of the Battle of Long Island, one of the American officers, Colonel Benjamin Tallmadge said:[1]

In the history of warfare, I do not recollect a more fortunate retreat. After all, the providential appearance of the fog saved a part of our army from being captured and certainly myself among others who formed the rear guard.

Following the Battle of Long Island, George Washington was in a fog himself as to what the British would do next. There were two possibilities: they could attack the Americans on the island of Manhattan, or they could sail up the Sound and land in Westchester County to outflank the colonists.

Nathan Hale

In an attempt to find out exactly what the enemy would do, Washington sent a spy into the British ranks. Nathan Hale, a young graduate of Yale, volunteered for this risky assignment. Dressed as a Dutch schoolmaster, the intrepid patriot made his way among the British in New York City and on Long Island, making notes in Latin which he hid in his shoes. In September 1776 Hale crossed the Sound from Norwalk, Connecticut, to Huntington on the North Shore of Long Island. Stephen Hempstead, a sergeant in Hale's company recounted what happened as follows:[2]

We left our camp on Harlem Heights with the intention of crossing over the first opportunity, but none offered until we arrived at Norwalk. . . . In that harbor there was an armed sloop and one or two row galleys. Capt. Hale had a general order to all armed vessels to take him to any place he should designate. He was set across the Sound in the sloop at Huntington (Long Island) by Capt. Pond who commanded the vessel. Capt. Hale had changed his uniform for a plain suit of citizens brown clothes, with a round, broad-rimmed hat . . . leaving all his other clothes, commission, public and private papers, with me and also his silver shoe buckles . . . and retaining nothing but his college diploma. Thus equipped we parted for the last time in life.

Hale soon thereafter fell into enemy hands. When he was brought to headquarters, General Howe, the British commander in New York, was so impressed by the diagrams of British forts found in the young man's shoes that he offered Hale his life on condition that he join the British army. Rejecting the offer, Hale was hanged as a spy. Before dying he made himself immortal by uttering the now famous words: "I regret that I have but one life to give for my country."

20

Retribution: Major André

Just as the Americans had their spies in the Sound area, so did the British, whose intelligence service maintained a network of informers. A center of their activities was Raynham Hall in Oyster Bay, the home of the Townsend family, where British Colonel Simcoe, commander of the Queen's Rangers, had his headquarters during the war. Little did he realize that there lived an American counterspy right in the house. She was Sarah Townsend, whose brother, disguised as a merchant, was residing in New York and acting as an agent for Washington's army. Sarah could not help noticing visits of Major André, an officer of the American forces, a traitor to the patriot cause now serving as a British agent. The alert girl noticed André taking a letter out of the cupboard which served as a drop for exchanging information. She immediately notified her brother. Entrusting her message to an unsuspecting Loyalist eager to win Sarah's affections, she managed, by ordering a special kind of tea from her brother's store, to get the information over to Manhattan without arousing any suspicion. Robert Townsend forwarded the message as soon as he could to Colonel Benjamin Tallmadge on the mainland. André's fate was now sealed. Tallmadge happened to be Hale's classmate and good friend at Yale, and the uneven balance between the two camps caused by Hale's execution could be evened out now in the ledger of fate. Tallmadge, on the basis of Sarah Townsend's message, could prove the guilt of André, who was then, in accordance with the custom of the time, executed as a spy. Although the traitor's death could not bring back to life his beloved classmate, Tallmadge could sooth the wound with the salve of retribution.

The Refugees

Retaliation, however, could not be exacted for all the suffering endured by the patriots during the war. With the enemy occupying Long Island, many of the inhabitants fled across the Sound to Connecticut as the war progressed. Theirs was the common fate of refugees. Having fled with few material goods, they found themselves in dire financial straits on the mainland. Time and again they would request permission from the committees of safety set up to aid the American war effort to return to their former homes on Long Island and bring back some of their possessions. Many returned empty-handed, telling sad stories of how their homes had been looted in their absence. Witnessing these atrocities the remaining Long Island residents who supported independence chose to stay in their homes rather than seek refuge on the mainland.

The terror of war struck also in Westchester County, where coastal residents, harassed by frequent raids from the Sound, fled farther inland to seek security in distance from the seashore. In the fall of 1776 when the British landed in the southern corner of Westchester County at Throgs Neck, a more serious threat arose: permanent enemy occupation. General Washington himself, in a letter to Congress on October 13, 1776, described the seriousness of the threat in these words:[3]

Yesterday the enemy landed at Frog's Point, about nine miles from here, farther up the Sound. Their number we cannot ascertain, as they have not advanced from the point, which is a kind of island; but the water that surrounds it is fordable at low tide. . . . From the great number of sloops, schooners, and nine ships that went up the sound in the evening full of men and from the information of the two deserters who came over last night, I have reason to believe that the greatest part of their army has moved upwards, or is about to do it, pursuing their original plan of getting in our rear. . . .

On October 18th, five days after they stepped ashore at Throgs Neck (Washington's "Frog's Point"), additional British forces landed at Pell's Point. Marching inland, Colonel John Glover, who had helped Washington across the East River after the Battle of Long Island, blocked their advance in a series of sharp engagements known as the Battle of Pell's Point.

The Battle of Pell's Point

Save this encounter, no major military action had taken place on the shores of Long Island Sound throughout history. Its significance lay not in the number of men involved, for only about 4,000 on the British side faced about 750 Americans, but rather is indicated in Washington's report to Congress: "Getting in our rear." If they had succeeded, the British might have destroyed the entire American army.

To understand the meaning of the original British plan, it is necessary to take a quick look at the strategic situation. After the Battle of Long Island, Washington evacuated his troops to Manhattan. Retreating later to Harlem Heights, he settled down in heavily fortified positions. Surrounded by the Hudson and Harlem rivers, Washington was in a defensible position unless Sir William Howe, a disciple of the formidable General Wolfe of Plains of Abraham fame, came around his flank and attacked him in the rear.

The enemy commanded the water routes on the Hudson as well as on the East River and the Sound. Pinning the Americans down on the Harlem Heights by an adequate besieging force, the British, through an amphibious operation, could threaten Washington's lines of supply running north through Westchester County to New England. The British could accomplish such an encircling move by landing at a suitable place on the Sound and marching westward to the Hudson—a distance of no more than six miles. By cutting the Boston Post Road and all other roads coming down from the mainland to Manhattan, they could have held back from Washington all sources of sustenance. This was the plan the British commander prepared, excellent in concept but very faulty in execution. To begin with, General Howe's intelligence reports left much to be desired when they selected the first landing place. The attempt at Throgs Neck placed the landing party on a section of the shoreline "which is a kind of island,"[4] as Washington put it—at high tide that is. At low tide salt marshes, difficult for troops under enemy fire to wade through, separated it from higher ground. After days of hesitation, enough for Washington to realize the enemy's intentions, the British moved their troops from Throgs Neck to Pell's Point. From here a good road led into the interior.

The landing party arrived on October 18th at the break of dawn. On the same day Washington began to fall back from Harlem Heights to avoid being trapped. The retreat was slow because of lack of adequate transports; those available had to struggle ahead on poor roads. It took Washington four days to reach new positions on the heights of White Plains in Westchester County, a distance of only eighteen miles. As this slow caravan was trekking northward, a flank attack by the enemy would have produced a devastating effect. But the enemy, now trying to carry out his design with great determination, met with an equally determined American opponent who stood in the way, who barred quick access to the interior, and who with amazing resourcefulness and courage defeated General Howe's encircling move.

The hero of this remarkable feat, Colonel John Glover, commander of an elite brigade from Massachusetts, conceived of a brilliant plan to delay the enemy's advance as long as possible in spite of overwhelming odds. He had 750 men facing 4,000 seasoned professionals of the invading force, three-fourths of them Hessians. Scanning the terrain, Glover realized that the many stone walls fencing off one field from another were excellent natural breastworks for an army on the defensive. Knowing that his assailants would rely mainly on cold steel, as they always had so far, he instructed his men to take up positions behind the stone walls in order to be protected against the initial volleys which usually precede the bayonet attack. Then, and this was the crux of the matter, he ordered the men to wait until the enemy would come close enough so that even the poorest marksman could not fail to hit the target.

Colonel Glover, remembering Bunker Hill, knew well what effect such delayed fire would have on the closed ranks of the assailants. He added something new to these tactics by staggering the lines of defense in a checkerboard fashion. While one section of his small force would take a stand on the right side of the road, another, a few hundred feet farther behind, would man the stone ramparts on the left side. Thus, when the first section delivered its deadly volleys and scattered the enemy, it was ordered to fall back without waiting for the assailant to reorganize and renew the charge. This way a fresh line of defenders was waiting for the next onslaught, with the same dire consequences to the attackers.

This remarkable engagement lasted the major part of the day. At the end of the battle, Glover still had 150 of his men and three fieldpieces in reserve which had not been involved in the fighting. The enemy suffered from 800 to 1,000 casualties, was totally exhausted and its morale broken. No further attempts were made to press the issue, and General Howe after marching to New Rochelle where he rested his troops for four days, finally decided on October 28th to attack General Washington's main force, by now well entrenched on the heights of White Plains. Ten days, in other words, had passed since the landing at Pell's Point, mainly because of Colonel Glover's remarkable tactics.

Some historians consider the Battle of Pell's Point one of the most decisive engagements of the entire Revolutionary War. There is no reason to dispute this interpretation except perhaps to add some qualification. If it was not one of the most decisive battles, it was certainly one of the most brilliantly conceived engagements. The many stone fences crisscrossing the terrain made it impossible for the British army to leave the road and march cross-country. The troops were forced to advance on the road leaving it only to attack Glover's men posted behind the stone fences. Each time heaps of casualties marked the attempt while the defender got away with a minimum of losses. It was like little Thermopylaes steadily moving backward, each time taking their grievous toll of General Howe's "Persians."

Shortly after the Battle of Pell's Point, the Americans surprised a detachment of British troops at Mamaroneck. In the ensuing battle on Heathcote Hill overlooking the Sound, the invaders suffered enough damage to prevent them from going on to White Plains where the main force of Washington managed to withstand several British attacks in late October, 1776. The withdrawal of the American army continued across the Hudson River, and henceforth the focus of activity in the Revolutionary War shifted to New Jersey and then to the South. Throughout the remainder of the war, however, Long Island Sound and the communities around it were the scene of important military activities.

Bushnell's Submarine

Sometimes these activities were more dramatic than successful, as was the case with the *Turtle,* the first submarine, invented by a young Yale graduate, David Bushnell. Major General Heath described Bushnell's invention as follows:[5]

This machine was worked under water. It conveyed a magazine of powder which was to be fixed under the keel of the ship, then freed from the machine, and left with clockwork going, which was to produce fire when the machine had got under way.

Bushnell had built his submersible near his home in Saybrook, Connecticut, and then brought it down the Sound to New Rochelle. It was a small device made of wood, capable of carrying one man under water who could propel it with a hand-driven screw. Weighing little, it could be lifted on board a ship or transported overland by wagon. In order to avoid the enemy, its inventor moved it first to New Rochelle and then across Westchester to the Hudson, whence it could be easily brought down to New York harbor, which teemed with suitable targets. In 1776 an attempt was made to blow up the British ship *Eagle,* but the volunteer who drove the "sub" could not attach the explosive charge to the metal-protected hull of its intended victim. Without being firmly fastened, the freely floating charge, after explosion, produced only a huge spray of water near the target without causing any harm to it. The *Eagle,* suspicious of the strange incident, merely left the site and cast anchor farther out in the harbor.

Bushnell's submarine made another try in the following year against the British frigate *Cerberus* off New London. Before it could approach its target, it was observed by the crew of a passing British schooner. Seeing the mysterious "fishing line" floating in the water, they decided to investigate and lifted the whole device, including the charge, aboard. The ensuing explosion demolished the vessel.

The War on the Sound

After the Battle of Long Island, the British sent detachments to occupy the rest of the island, to give support to the Loyalists and force the rebels into submission. Atrocities of all kinds were committed against the latter in which the Hessians particularly distinguished themselves. Requisitioning for His Majesty's armies, however, was carried out on a bipartisan basis—that is, against patriots as well as Loyalists.

American Raids

The situation turned from bad to worse when the patriots began to organize their forces on the mainland. They carried on a systematic war of raids across the Sound as privateers, capturing not only ships of the British fleet but vessels engaged in trade with the enemy. The Sound thus became a theater of war, its silent waters often carrying boats sliding swiftly under the cover of night to harass the enemy and carry out bold raids against his supplies, ships, or forts on the North Shore. Rowboats used in the whaling industry were often employed for such purposes.

Off City Island, for example, a small group of whaleboat men from Darien, after portaging their craft over Rodman's Neck, drew up alongside a merchant sloop known to be a regular supplier of the British ship *Schuldham.* Taking over the sloop and pretending to be its regular crew, the whaleboat men were allowed to board the *Schuldham* in the early hours of the morning. Needless to say, the British crew on board had a rather rude awakening as they saw their captain, with a gun at his head, guiding the vessel eastward to Stamford. Patriots throughout the Sound were thrilled to hear about this exploit. They remembered the raiding parties frequently put ashore by the *Schuldham* to requisition supplies. The inhabitants of Westchester remembered even more the brutalities perpetrated by the raiders as they took away their provisions, leaving behind much grief and resentment.

Kidnappings

Of course, Westchester patriots were by no means the only people to suffer indignities. Coastal Connecticut was also a favorite target of British raiding parties. Crossing from British-held Long Island, the enemy picked up all manner of things from farms and villages, not just material goods. Sometimes the objective of a raid was to capture a prominent person who could then be exchanged for someone who had been taken by the Americans. A classic example of this little game practiced with great relish by both sides involved General Gold Selleck Silliman of Fairfield, Connecticut. Loyalist whaleboat men crossing the Sound from Lloyd's Neck in the spring of 1779 captured him in his home. Witnessing the incident, the lady of the house recorded in her personal journal:[6]

At a midnight hour when we were all asleep, the house was attacked. I was first awakened by the General's calling out, 'Who's there?' At that instant there was a banging at both doors, they intending to break them down or burst them open. . . . My dear companion then sprang up, caught his gun and ran to the front of the house and, as the moon shone brightly, saw them through the window and attempted to fire, but his gun only flashed and missed fire. At that instant the enemy burst in a window, sash and all, jumped in, seized him and said he was their prisoner, and must go with them.

The British, after holding their captive for a year, exchanged General Silliman for a prominent Long Island Loyalist, Judge Thomas Jones, caught off guard by the patriots at a dance in November 1779, taken prisoner and brought across the Sound to Connecticut. For a while the judge was confined in, of all places, the home of General Silliman, whose wife, a perfect hostess, saw to it that he was well treated and well fed. Perhaps it was due to Mrs. Silliman's generosity that, when the time came in the spring of 1780 to exchange the prisoners, the arrangements bore no trace of resentment or hostility. Meeting in their respective boats in the middle of the Sound, Silliman and Jones first dined with each other before continuing their journey.

Military Raids

The amicable relationship existing between Judge Jones and General Silliman should not be considered typical of the attitude of Loyalists toward patriots and vice versa. Usually there was great animosity on both sides, and this feeling frequently manifested itself in cross-Sound expeditions aimed at doing as much damage as possible to the enemy. The raid launched in May 1777 to destroy the hay stored by the British at Sag Harbor may be taken as a typical example. The party under the command of Colonel Return Jonathan Meigs set out in whaleboats from New Haven. Then they went north to Sachem's Head Harbor, Guilford, where the Sound is narrower, and from there they crossed to Southold. Working their way to Sag Harbor in utmost secrecy, they surprised the British. Although fired upon for almost an hour, the Americans suffered no casualties; yet they managed to destroy a considerable quantity of British provisions. In appreciation for his outstanding work in leading the expedition, Congress subsequently presented a sword to Colonel Meigs.

Several months after the Sag Harbor expedition, in August 1777, the Americans once again crossed the Sound, this time to attack the Queens County Loyalists who held the Presbyterian meeting house in Setauket. Not much could be accomplished this time by way of positive results. After capturing a small quantity of supplies, the raiders had to break off the expedition and flee back across the Sound, where a British fleet had appeared that could have cut off their return to the mainland or could have sent a landing party ashore to help the besieged Loyalists. Undaunted by this failure, General Samuel Holden Parsons, leader of the expedition, ordered the Americans, after safely returning to Connecticut, to try again.

Toward the end of 1777 he conceived of a daring plan to launch a three-pronged attack on Long Island from the mainland. One expedition was to cross the Sound from Saw Pitts (Port Chester, N.Y.) to Hempstead, another to attack Huntington, and the third, under the command of General Parsons, was to land on the eastern end of Long Island. He selected December 9th as the date for the invasion, but rough weather precluded a crossing at Saw Pitts. The expedition had to be canceled, but the other two groups left from Norwalk. The party headed for Huntington was intercepted by a British sloop sailing eastward on the Sound from New York to Newport. General Parsons, however, managed to reach the eastern end of Long Island, where he captured British supplies.

In September 1778 the Americans tried again. This time their objective was Fort Franklin on Lloyd's Neck, built on the high ground overlooking the entrance to Cold Spring Harbor. It was named in honor of Sir William Franklin, Benjamin Franklin's Loyalist son, the last royal governor of the New Jersey colony. The expedition, over one hundred men strong and led by Major Tallmadge, set sail from Shippan Point, Stamford. Unable to take the fort, Tallmadge nevertheless captured prisoners, bringing them across the Sound to Connecticut. This was not the only attempt. In July 1781, after France had become an active ally of the colonists, French and American ships joined forces on the Sound to take Fort Franklin, but they lacked the heavy artillery necessary to reduce the defenses.

Three monts after this futile effort, the Americans attacked Fort Salonga in Huntington. Crossing at the mouth of the Saugatuck River in Connecticut, they landed west of the fort, and the following night, guided by a local patriot, launched a successful attack. The American victory gladdened the hearts of Long Island patriots, because the British forts along the North Shore were a source of constant harassment to the inhabitants. As late as November 1782, when peace negotiations were already going on in Paris, the British, who at the time still occupied Long Island, forced the people of Huntington to help them build Fort Golgotha. Pressing the population into service was bad enough, but to add insult to injury the British selected as the site for their fort the town cemetery. Enraged as they were at this arrogance, North Shore residents could do little to retaliate except to tear down the fort as soon as the British withdrew after the end of the war.

Prior to the conclusion of hostilities, however, the Americans had several opportunities to inflict serious damage upon the enemy. In November 1780 Colonel Tallmadge, setting out from Fairfield, Connecticut, invaded the Mount Sinai area on the North Shore. At first impeded by heavy rains, he pressed on as soon as the weather cleared. After capturing Fort George he went on to Coram, where he destroyed several hundred tons of hay stored by the enemy.

On December 5 1782 an expedition was to start from Shippan Point, Stamford, to carry out a raid against Huntington under the leadership of Colonel Tallmadge. Heavy storms prevented him from crossing the Sound. Two days later Captain Caleb Brewster, who in the meantime had discovered three enemy vessels off Huntington, decided to prepare another attack. During the fighting he suffered a heavy wound in the chest caused by a bullet fired from close range. Unable to lead his men, the brave captain had to call off the raid, known as the boat fight. After recovering from his severe wound, he led another daring expedition in March 1783 against the British vessel *Fox* stationed off Fairfield, Connecticut. This time he succeeded so well that none of his brave band suffered any casualties, and the captain himself lived on to the ripe old age of eighty.

Degeneration of the Raids

Aside from these major raids led by regularly appointed officers of Washington's army, there were innumerable smaller excursions across the Sound led by different types of people. At the beginning of the war, it was a patriotic thing to organize parties ready to cross the Sound in whaleboats. They would row silently into a cove or inlet, hide the boat at a convenient place, and then march inland to carry out various designs. Human nature being what it is, the whaleboat men in time forgot the cause they were supposed to serve and looked about for hidden treasures or other kinds of easy loot. More often than not toward the end of the war they practiced outright robbery under the guise of harassing the enemy. The enemy might be anyone who had money or movable possessions.

"Whaleboating" went on in both directions—north and south. Smugglers, profiteers, spies, and speculators would go to the mainland, each pursuing his special aim. A lucrative form of profiteering was to buy goods in New York City not available in rebel-held territory, ship them across the Sound at night, and barter them for provisions on the mainland. These could be sold at high prices to the British. But it was an involved and rather risky business, strictly forbidden by the authorities on both sides yet practiced by many daring men, since the British, quartering their troops in the villages of Long Island, were badly in need of provisions, especially in the winter months. They paid good prices to the local farmers, too, who on the other hand could not spend their money very well, normal trade being at a standstill. Law and order at the same time broke down, and robbery, accompanied by all sorts of violence, grew rampant as the war went on. The whaleboat men, realizing that there was much money held by the farmers, began to pay nightly visits to them, breaking down doors and forcing them to disclose the hiding places where they stashed away their hoards. Years later construction workers would come upon such "pots of gold" left in the hiding place by the owner who died a sudden or violent death before he could reveal his secret.

British Raids

In order to terrorize the population living on the mainland shore of the Sound, the British carried out systematic raids of large-scale devastation. Little aid could be given to the towns lining the seacoast in Connecticut, and the patriots found themselves literally at the mercy of the enemy. In February 1776 a British force under the command of General Tryon, the former governor of the New York Colony, appeared in the Sound off New Rochelle. His objective was the saltworks in Greenwich, Connecticut, the principal source of that all-important commodity for the Continental army. When the news of an impending attack reached Greenwich, General Israel Putnam gathered his men and marched out to meet the enemy. Forced to retreat, Putnam decided to head for Stamford, where he might be able to obtain reinforcements. Being pursued at close range, he decided to take a little detour down the side of a steep hill at full gallop. With this daring ride he succeeded in throwing the British off his trail. His pursuers, after firing a few shots which missed the general, abandoned the chase. "Old Put," as he was affectionately called, made good his escape.

Three years later, with forty ships under his command, General Tryon decided to sail up to New Haven, hoping to accomplish even greater feats against a much weaker opponent. The appearance of such a large fleet in the harbor alarmed the townspeople, among them Ezra Stiles, president of Yale College, who watched the British fleet through a spyglass from atop the college chapel. Stiles promptly packed up his daughter and the college records and sent them away for safekeeping. As news arrived of enemy contingents landing in East and West Haven, members of the militia, among them many students at Yale, quickly mobilized to meet the invader. The commander of the western wing of the British striking force, General George Garth, realizing that the Americans were ready to make a determined stand, changed the route of his march to the city in the hope of avoiding a direct encounter. But it was to no avail, because the irate patriots continuously harassed the British, firing upon them seemingly from nowhere. When the troops finally reached New Haven, they were so exasperated from the losses they had suffered on the march that they let loose with a reign of terror. Robbery, rape, and murder were the order of the day. The unruly British soldiers even attacked their own commanding officers who made efforts to restrain them. A British adjutant by the name of Campbell was shot to death by one of his own men when he tried to identify the soldier in his company who had violently wrenched a gold ring from the finger of a New Haven woman. While all this was taking place, Tryon had destroyed Fort Hale on the peninsula of East Haven. He did, however, run into such spirited opposition from the Americans that both he and Garth retreated to their boats in the Sound the day after they landed. Many of their men were suffering from hangovers following the celebrations staged after the reign of terror.

Three days after the invasion of New Haven, the British had recovered sufficiently to stage a similar attack upon Fairfield. Repeating the same pincers tactic, Garth approached from the west while Tryon marched from the east. Plunder, rape, and murder were once again the order of the day. The defenders conducted themselves well and managed to hold a small fort against the onslaught from a British galley. Just as in New Haven, the American practice of sniping at the British whenever and wherever they could aroused the wrath of the enemy so that in retaliation they burned the town. Tryon justified this act by saying that some of the homes which were set afire harbored snipers. William Wheeler, an American eyewitness to the destruction of the town, looked at the burning of Fairfield in a different way, however, as he related in his diary:[7]

At night the British placed guards round the town which were plainly seen by the burning houses, the while many a column of fire from the flaming buildings and frequent flashes of lightning from a western cloud with discharges of cannon and musquetry formed a prospect the most gloomy and comfortless imaginable to the poor inhabitants who, many of them, sheltered only by the canopy of heaven, without a second suit to their backs, or a penny in their purse, behold from a distance the fruit of all their toil and labor expiring in a cloud of smoke and cinders.

Following the attack on Fairfield, Tryon crossed the Sound to Huntington and took on supplies in order to continue the devastations of the mainland towns. Recrossing the Sound on July 10th, he again divided his troops to attack his next target, Norwalk, from both sides, with General Garth coming up from South Norwalk while Tryon approached from the east. As at Fairfield, he also included the burning of the town in his strategy. The story is told how Tryon sat à la Nero in a comfortable chair on top of a hill watching the conflagration.

But for General Washington's prompt decision to send General Heath to the Sound, Tryon's next target might have been Stamford. Tryon, looking not for serious encounters but for easy victories, abandoned the plan, and lucky Stamford was spared the visit of his looting and burning mercenaries.

The Sack of New London

In September 1781, as the Americans and their French allies approached Yorktown, Virginia, for what would be the last battle of the Revolutionary War, more than two dozen British ships carrying several thousand men appeared in the Sound off New London. The expedition was led by Benedict Arnold, who earlier in the war had gone over from the American to the British side. Arnold caught New London by surprise because, despite the fact that there were fortifications at New London and Groton, not much had happened in this area. Thus, when the British attacked the two forts on September 6th, the Americans were caught off guard. The defenders of Fort Trumbull managed, however, to spike the guns of the fort before going over in small boats, one of which was captured by the enemy, to defend Fort Griswold. While this was taking place, Benedict Arnold launched an attack upon New London itself. The patriots rose to defend the town, and although Arnold contended in his official report that the opposition was fierce, the Americans were really at a disadvantage. As had happened elsewhere along the Sound coast of Connecticut, buildings were set afire. In his memoirs General Heath described the scene as follows:[8]

Arnold himself continued on the New London side, and while his troops were plundering and burning, was said to have been at a house where he was treated very politely; that while he was sitting with the gentleman regaling himself, the latter observed that he hoped his house and property would be safe; he was answered that while he [Arnold] was there, it would not be touched; but the house, except the room in which they were, was soon plundered and found to be on fire.

After subjecting the town of New London to the horrors of war, the British concentrated their attack on Fort Griswold. Arnold, watching the operation from the top of a hill, at first thought that the siege might be very costly to his forces and sent a message to the troops who were to attack the American stronghold to be prepared for hard fighting. But Arnold misjudged the situation. The fort fell before the onslaught, and Arnold, elated by the unexpected success, sent a grandiloquent report to headquarters about the engagement: "The attack was judicious and spirited and reflects the highest honor on the officers of the troops engaged, who seemed to vie with each other in being the first in danger."[9] The officers and men may have fought with valor, but not all of them acted with "highest honor." When Colonel Ledyard, the American commander of the fort, offered his sword in sur-

31

render to his adversary, the British officer, feigning acceptance of the surrender, turned the sword around and plunged it into Ledyard's heart. The British then decided to blow up the magazine of the fort, but before doing so removed the American wounded, whom they dumped one on top of the other into a wagon. Losing control of the heavily loaded vehicle on top of a steep hill, they simply let it go, with the result that some of the wounded men were killed in the wreck. As a final grisly gesture, before withdrawing from the New London area, the British burned the nearby village of Groton.

The War of 1812

When the Revolutionary War was at last over, the people of New London undoubtedly breathed a huge sigh of relief as they rebuilt their town and resumed the seafaring activities which constituted their livelihood. Up and down the mainland side of the Sound as well as on the North Shore of Long Island, the economic situation with its heavy emphasis upon fishing, trading, and shipping, returned to normal after the war, but normalcy did not endure forever. When war broke out between the Americans and the British in 1812, the Sound again became the target of the enemy, although not immediately.

It appeared at first that the Sound would be spared the depredations of war, since New Englanders strongly opposed hostilities against England. Their main interest was trade and they could not see any advantage in having their means of livelihood cut off by the enemy and their coastal areas, including the Sound, exposed to his destructive raids. Aware of this hostile sentiment toward the war, the British at first were cautious not to reverse the antiwar feeling in New England.

As time passed and hostilities went on without an end in sight, Great Britain decided to blockade the Sound. Trade stopped, and, worse still, the British captured a number of ships owned by Long Islanders. If vessels were seized while crossing the Sound, the enemy either destroyed them or ransomed them at exorbitant figures. This buccaneering sanctioned by war continued while hostilities lasted, arousing bitterness on both sides of the Sound. Just as in the Revolutionary War, opposition to the British brought many volunteers to the American side from various parts of the Sound. Those who came from Mystic, Connecticut, had a special reputation for hating the British, and with good reason. They had suffered heavy losses from the enemy's random destruction of their ships. In a successful retaliatory effort the townspeople sent a galley called the *Yankee* out onto the Sound where it captured several British ships in the vicinity of Plum and Fishers islands.

Across the Sound on the eastern end of Long Island, the threat of British attack prevented citizens from venturing out onto the open waters. Instead, they stayed put, in order to defend their homes. In fact, they were ordered to do so by the governor of New York because of the increasingly dangerous situation on the north fork of Long Island. British ships sent raiding parties to seize cattle belonging to the residents. There were also several battles fought between the Americans and the British. One occurred at Riverhead in June 1814 when the Americans learned that two British barges were out in the Sound. A group under Captain Terry Reeve gathered on the beach as approximately fifty British soldiers proceeded toward the shore. They apparently intended to take an American ship near the beach, but as they came close to shore the American militia fired upon them and sent them scurrying.

Elsewhere the Americans were not so successful. On Gardiner's Island, for example, the British landed and seized cattle and other supplies, while on Block Island they managed to gain complete control after sailing through the Sound from west to east. The British then established their blockade of the Sound, which the American Commodore Stephen Decatur attempted to break, only to be hemmed in at New London by the enemy. When this occurred, the Connecticut militia was ordered to New London to defend the city against an imminent attack. At the same time the women and children of New London were moved to Norwich. Decatur even had to move his ships into the Thames River away from the Sound. Growing restless as time went on, he attempted to bolt the British blockade and make his way out. This proved to be impossible because the British somehow learned of Decatur's secret escape plan. The commodore and his men went to New York by land, leaving the fleet on the Thames.

Farther south along the Sound coast of Connecticut, the British did considerable damage. Near Saybrook they demolished ships, while in Stonington they bombarded the town, setting fire to a number of buildings. One saving grace, however, was the fact that the British flagship with its deep draft, could not approach the town but had to stay a mile and a half out in the Sound. The British bombarded Stonington for two days and then sailed away. In March 1815 they left the Sound altogether, leaving behind much bitterness among the residents. At times during the war bitterness was also caused by interference with American trade by the British privateer *Liverpool Packet*, which cruised the Sound preying upon American commerce. Retaliation was not always as successful as in the fall of 1813 when a torpedo vessel created out of an American schooner was captured by the British. The boat exploded while it was in enemy hands, killing almost a dozen British soldiers. Exploits such as this may explain why the United States government decided after the war to build a fort for the protection of the western entrance to the Sound. The site selected for the imposing new Fort Schuyler was Throgs Neck. Fortunately, no need ever arose for the fort to carry out its mission. After 1814 the Sound never saw a flag flying atop a hostile vessel in war.

The steamboat "Fulton" traveled from New York to New Haven in 1815 (Samuel Ward Stanton, "American Steam Vessels")

3

Steam Comes to the Sound

YOUNG PEOPLE IN THEIR LITTLE SUNFISH SAILBOATS OR FIBERGLASS POWERBOATS out on a summer's day when Long Island Sound is dotted with cheerful white sails would be surprised to see a picturesque side wheeler plowing through the waters. To the young of today such an antiquated craft used for transporting passengers would seem out of place on the Sound, which to them is nature's gift to be enjoyed mainly as an aquatic playground. They would perhaps be equally surprised to hear that there was a time when graceful steamboats plied the Sound in a constant procession from New York City to towns along the shore, even to points as far away as Boston.

The Era of Steamboats

Yet it is easy to comprehend the past. The future is enigmatic. Many years ago captains of the sailing vessels who caught their first glimpse of a steamboat on the Sound could not understand the meaning of what they saw: a small, smoke-belching, noisy contraption churning the waters with its huge side wheels as it made its way slowly eastward. It was the *Fulton* trying in the year 1815 to make the seventy-five-mile trip from New York City to New Haven. Perhaps these incredulous captains and surprised spectators had heard of the man whose name they saw inscribed with large letters on the ship's side and of his epoch-making journey in 1807 when he steered his little *Clermont* from New York to Albany. Fulton, to be sure, was not the first to experiment with steam, but he was one of the earliest to harness successfully this new form of energy for transportation on rivers and inland seas.

35

For a while delayed by the War of 1812, steamboating in the postwar period became a big business, altering the old ways of water transportation on Long Island Sound. It was not all smooth sailing for the steamboats, however, for unlike the placid Hudson the Sound could whip up angry waves in a storm. Furthermore, steamboat captains on the Sound had to contend with fog and, in winter, with floating ice. Moreover, there were treacherous tides at both ends of the waterway, at Hell Gate in the west and at the Race near Block Island in the east.

The first attempt to navigate the Sound with a steamboat took place, as we have seen, in 1815. The *Fulton* succeeded in reaching New Haven in eleven hours. The return trip, postponed for a day because of bad weather, took fifteen hours. It was, nevertheless, quite an accomplishment for her captain Elihu Bunker, who had to contend with rough seas, wind, and fog. Undaunted, he pressed on to achieve a monumental first in the history of Long Island Sound. Commenting on this feat, the *New York Evening Post* declared:[1]

The facility with which she [the *Fulton*] passed Hell Gate . . . surprised everybody on board, and satisfied them that no vessel can be so well calculated to navigate the dangerous channel as the steamboat. . . . The boat passing these whirlpools with rapidity while the angry waves are foaming against her bows, and appear to raise themselves in obstinate resistance to her passage, is a proud triumph of human ingenuity.

The year after this momentous voyage, Captain Bunker assumed command of the steamer *Connecticut,* a gleaming new vessel originally built for the tsar of Russia and supposed to bear the name *Emperor Alexander.* Instead, she wound up on the Sound to thrill thousands of passengers and onlookers from both shores who were quite impressed by this white queen of the waves with her bright green trim.

The success of the *Fulton* and the *Connecticut* opened up a new chapter of transportation. Regular steamship service with two weekly round trips, soon to be extended to three, between New York on the one hand and New Haven and New London on the other was begun in 1817. One of the first travelers attracted by the new service was President James Monroe. Shortly after his inauguration in March 1817, he made a tour of the New England states and en route sailed through the sound on the *Connecticut.* The railroads, arriving on the scene some years later, added a new element of speed to this service. Connecting such inland cities as Hartford and Springfield with New Haven and coordinating their schedules with the steamships, these trains added a new combination of comfort and speed, and soon extended their runs to Boston. The old combination of sailboat and stagecoach suddenly became a relic of the past. The New Haven Railroad, of which we will hear much more later, ran trains from Hartford to New Haven, leaving at 5:30 A.M. to make connection with the 8 A.M. steamboat to New York. If train and boat were on time, the passenger could expect to arrive in New York at 1:30 P.M., a trip which in the old days of sail and stage took two days.

36

Breaking the Fulton Monopoly

Further rapid developments would have followed had it not been for the Fulton interests' monopoly of rights to operate steamboats on the waters of New York State. In 1825 Captain Benjamin Beecher devised a clever scheme to beat this monopoly. Using a schooner to tow his newly purchased steamboat the *United States* from New York to Byram Cove in East Port Chester, the most westerly point along the Sound shore of Connecticut, he observed the provisions of the monopoly. Once across the state line, he got up steam and the *United States* proceeded on her journey. On the return trip Beecher's steamboat had to dock at Byram Cove, where passengers transferred to stagecoaches for the rest of the trip. This situation persisted until the supreme Court ruled in the famous case of Gibbons vs. Ogden that the steamboat monopoly was unconstitutional.

The decision must have gladdened the heart of many a weary traveler, because the overland route was anything but comfortable. There were dreadful ruts along the Boston Post Road, a condition resulting from the unwillingness of coastal towns to appropriate funds for the repair and maintenance of a road considered unnecessary in an age when most people used the Sound as a highway. This age of the slow sailboat and the bumpy stagecoach now came to an end. Steamboat travel offered fast and even luxurious service. By the late 1820s it lured passengers with such amenities as carpeted and curtained salons and on some of the boats libraries containing hundreds of volumes.

The Golden Age of Steam Navigation

Following the Supreme Court's ruling against the Fulton monopoly, a fierce competition arose among rival steamship companies resulting in many improvements to attract passengers and freights. Low fares, luxurious accommodations, safety and, before everything else, speed were offered to increase prestige and through prestige business. The golden age of steam navigation opened on the "American Mediterranean" with fanfare, with spectacular speed races, and with some dreadful accidents. Its beginnings were slow and laborious, however. One of the principal difficulties was fuel, which consisted of cordwood of good dry pine. It burned well and fast, but a lot of it had to be carried on a long trip. When the *Clermont* made its memorable trip on the Hudson it was greeted with derisive laughter because its deck was loaded with "pine sticks," leaving little room for cargo. The captains of cargo-carrying sloops could not understand the purpose of the steamboat. The significance of steam as the new source of power escaped them; they just could not see beyond their sails. They were right, though, in one respect: sails never exploded. Boilers did.

With the introduction of more powerful engines, the need for fuel increased. While the *Clermont* carried a dozen cord of wood, the fast steamers on the Sound needed so much fuel that they could not carry all of it. Instead of putting into a port to reload, they would pick up a wood sloop off Fishers Island and refuel without stopping. Competition forced the captains of the early Sound steamers to resort to such devices. Of course, coal would have occupied much less room, but coal firing was tried successfully only in 1836. After that it spread rapidly, replacing wood in a few years.

The Danger of Boiler Explosions

The other problem, and a very dangerous one, was the inability of the boilers to withstand sustained high pressure. The early engines operated with low-pressure steam, but the desire for greater speeds and more power on the rough waters of the Sound produced stronger engines. Few explosions would have occurred had it not been for the willingness of the captains to engage in races. Indeed, the history of early steamboat navigation is replete with dramatic stories from the Mississippi to Long Island Sound. One tragic case involved the side-wheeler *New England,* which exploded in Essex harbor on the Connecticut River, seven miles in from the Sound. The *New England* had just arrived after an exciting voyage from New York City. En route she met another steamboat, the *Providence,* engaging her in an impromptu race up the Sound. The *New England* won, but she had built up so much steam pressure that her boilers exploded, killing and injuring fifteen people. Commenting on the tragedy, the *New Haven Daily Herald* declared:[2]

This is the most terrible accident of the kind, probably, that has ever transpired, and the public anxiety is extremely alive to know from what cause it was produced. The boat was entirely new, built on the most approved model, and money and means were not spared to render her one of the best and safest boats upon our waters. Her boilers were of the best and stoutest copper and her commander an old and experienced captain. Some one must be responsible for a dreadful account.

Perhaps the answer lay in the fact that winning a spectacular race, even though the risks were great, meant not only fame and glory for the captain but better business for the company. Still, it was necessary to assure the passengers that if an explosion occurred they would not be killed, thrown overboard, or scalded to death. Various devices were therefore used to dissipate passenger anxieties. It was not uncommon, for example, at the beginning of the steamboat era to tie a steamer to a sailing vessel and have the passengers sit in the latter while the steamboat pulled it. There was still fear in some circles, however. Typical of the reaction to this newfangled invention was the feeling of a young boy who was on hand in Stamford in 1825 to greet the first steamboat which appeared in that Sound shore community. The boy declared:[3]

The *Oliver Wolcott* came up into the harbor. . . . The whole population of men and boys were there. A great crowd stepped on board as soon as she had made fast and one boy at least had a mortal fear of going near the boiler. In a moment the steam was let off, and such a fearful scream it made to our unaccustomed ears!

One method of alleviating the fears of steamboat passengers was to remove the boiler from the hold of the ship and raise it above the upper deck. Since explosions tend to go upward, the passengers, located on the lower deck, would feel more secure. But machinery, towering high above the deck, lent the ship an odd appearance. In the year 1842 the famous English writer Charles Dickens visited the United States and rode on a Sound steamer from New Haven to New York. Dickens was apparently startled by the appearance of the steamer *New York,* which carried him down the Sound, because he took the trouble to note that the main deck looked like a warehouse filled with caskets and crates. On top of it was the open promenade deck where the passengers could enjoy the view in good weather. Towering above this deck was part of the machinery, with the connecting rod working up and down like a top-sawyer. The helmsman sat in a little house in front of the two tall chimneys. The steering wheel, located in the little house, was connected with the rudder by chains which extended over the whole length of the deck. All the exposed equipment must have offended Dickens's aesthetic sense, because he remarked the ungainly appearance of not only the *New York* but of other steamships which they passed en route through the Sound.

Steamboat Racing

Another aspect of steamboating on the Sound absorbed the attention of both participants and spectators perhaps to an even greater degree than the strange appearance of the early steamboats. This was the rivalry of some of the most colorful characters of maritime history as they struggled to dominate the Sound. The man who made the greatest contribution to this brilliant period was Cornelius Vanderbilt. Although he was better known for his accomplishments in the railroad industry later in life, Vanderbilt began his business career as a boatman in New York harbor. After the elimination of the Fulton monopoly, he participated in various ventures on the Hudson as a ship owner and operator. At this time, in the early 1830s, he was attracted to the Sound by the rapidly increasing number of travelers between New York City and communities up and down the waterway but particularly between New York and Boston.

Since the opening of the Erie Canal in 1825, the port of New York had become a bustling commercial center, indeed the most important city on the Atlantic coast. Cornelius Vanderbilt decided to capitalize on New York's position by inaugurating a new service from that city to Boston. Happily for Vanderbilt, a railroad was being built between Providence and Boston, and it promised to reduce travel between those two cities to one-fourth of what it was by stagecoach. Vanderbilt would use this new service by running steamboats from New York City to

Providence, where the passengers would transfer to the railroad for the rest of the trip to Boston. In order to insure a swift and pleasant sea voyage through the Sound for his Providence-bound passengers, Vanderbilt ordered a magnificent new steamboat, the *Lexington*. The finest New Jersey white oak went into its construction to make it so sturdy that it could defy the elements with ease and inspire public confidence. The engine was of the latest design with boilers which could withstand the highest pressure when plowing the waters of the Sound at great speed. No explosions were to be feared on this luxury ship. The *Lexington* had an overall length of more than two hundred feet and gross weight of nearly five hundred tons. In comparison with Fulton's first ship, the *Clermont,* it was a real giant. There was indeed no similarity except that the *Lexington* too was to make history. On her maiden voyage in January 1835 the ship made the 210-mile run to Providence in twelve and a half hours, averaging seventeen miles per hour, an excellent record. It was not to last very long, however, because despite all of Vanderbilt's efforts to secure supremacy on the Sound, the competition would not resign itself to accepting his leadership. His rivals were that hardy type of boatmen of whom Vanderbilt himself was one. They were all products of the rough conditions of their trade, the final result of a long contest for the survival of the fittest on the waters of New York harbor, the Hudson, the New Jersey coast, and Long Island Sound. Soon, an opportunity to match their strength and ingenuity presented itself.

Two years after the opening of the Boston–Providence railroad, New York capitalists built a line from Providence to Stonington on the eastern edge of Connecticut, shortening the sea route between New York City and Providence. Boats landing in Stonington were spared the treacherous voyage around Point Judith, a rough corner on the western edge of Narragansett Bay. Navigation was especially hazardous there because the currents coming from the Sound, the bay, and the ocean created unpredictable conditions. Aside from this, trains starting from Stonington could reach Providence much faster than the boats. Passengers on the Boston run would certainly prefer this alternative.

This new development, however, would damage the reputation of Providence as a transfer point between ship and rail on the New York–Boston run. Providence had enjoyed this privileged position since the time of the Fulton monopoly. To combat that monopoly, the state of Connecticut had passed a law forbidding ships of New York State to put into Connecticut ports. Providence, being the nearest harbor beyond the forbidden waters, became the main destination of ships sailing from New York. Although the Supreme Court had ended the Fulton monopoly, in the meantime one monopoly had created another, and Providence businessmen, unwilling to relinquish their town's special position, now decided to challenge the company which used Stonington as its terminal point. They therefore ordered a new steamboat, the *John W. Richmond,* but the Stonington group managed to persuade Cornelius Vanderbilt to part with the *Lexington* for $60,000. Once each

40

FARE REDUCED.
ONLY THREE SHILLINGS TO NEW-YORK

NEW ARRANGEMENT
BETWEEN
MUSQUETO COVE, NEW-ROCHELLE, AND NEW-YORK.

THE Steam-boat LINNÆUS, CAPTAIN PECK, will leave Musqueto Cove every morning (Sundays excepted), at half-past 6 o'clock, New-Rochelle at half-past 7, and White Stone at half-past 8. Returning, leave New-York, from Fulton Market Slip, at 3 o'clock in the afternoon, and land passengers at the usual landing places.

STAGES will be in readiness to convey passengers to Flushing, Buckram, and White-Plains.

FERRY.—Persons wishing to cross to and from Long Island to the Main Shore, will do well to try this conveyance; it will be a saving of time to travellers, there being no ferry boat at this place.

Horses, carriages, &c. will be taken and put off at either side.
New-York, May 11

PRINTING PRESSES.

Early nineteenth century steamboats reckoned fare in shillings ("Standard-Star", October 8, 1934)

NEW ARRANGEMENTS
FOR
NEW ROCHELLE
FARE 12½ CTS.

THE STEAMBOAT
SUFFOLK
Capt. HOFFMIRE,

Will leave New-York from Fulton Slip, every Afternoon, (Sundays excepted,) at 4 o'clock.

Returning will leave New Rochelle every morning at 7 o'clock.

JUNE 16th, 1849.

Steamboat fares were relatively inexpensive in the nineteenth century ("Standard Star", October 8, 1934) group had secured a good ship, preparations were made for a race which was long remembered on the waters and shores of the Sound.

The race was scheduled to start in Stonington after both boats picked up New York–bound passengers from the train which had just arrived from Boston. In advance of the race the captain of the *Richmond* had selected the most resinous wood to fire his boat during the contest. Although the *Lexington* got off to a good start with heavy smoke billowing from her stacks and sparks flying, the *Richmond* swept through the Sound with pillars of fire rising from her stacks and reached Hell Gate an hour before the *Lexington*.

Tragic Accidents

Far less fortunate was the steamboat *Atlantic,* considered the largest and finest on the Sound at the time of her launching. On Thanksgiving Day 1846 in near-freezing temperature, a fire broke out aboard the proud vessel. The crew contained the flames, but the ship, having lost all power, was wrecked on the rocks at Fishers Island.

Similarly tragic was the fate of the *Lexington,* which ended her short but dramatic career only a few years after the famous race with the *Richmond.* The demise of the *Lexington* came on a freezing January night in 1840 after the ship had converted to coal firing to increase her efficiency. Coal firing required a forced draft provided by two blowers. The coal in the furnace burned so fiercely that the chimneys close to the deck turned red hot. Part of the cargo, cotton for the New England textile mills, piled close to the stacks, caught fire. The blaze spread rapidly, engulfing the entire ship. Captain George Child attempted to steer his steamer toward the North Shore of Long Island, but this proved impossible after the tiller ropes burned. The lifeboats, crowded with panicky passengers, were quickly swallowed up by the angry waters of the Sound. One of the lifeboats made it to shore at Setauket with only a coat aboard, whose owner, along with 119 others, had drowned in the icy waters. Only four people survived. The poet Henry Wadsworth Longfellow, who had booked passage on the *Lexington,* escaped the disaster because he had to stay in New York City for a lecture.

The *Lexington* disaster shocked the public. Nathaniel Currier commemorated the event with a print depicting the sinking of the *Lexington.* The picture proved exceedingly popular and helped make Currier's reputation even before he joined with Ives. Some of the cotton which formed the fatal cargo washed ashore and was made into *Lexington* shirts to be worn as mementos of the role which unburned bales of cotton played in the tragedy. Those who survived managed to stay alive by clinging to them. The story of one survivor was graphically related in the *Long Island Democrat:*[4]

Mr. Crowley, 2d mate of the *Lexington* drifted ashore near Riverhead, on Wednesday last, at nine o'clock, having been forty hours exposed to the severity of the weather, after which he made his way through large quantities of ice and snow, before gaining the beach, and then walked three-quarters of a mile to the house where he now is. His feet and hands are a little frozen.

The latter was quite an understatement in view of the fact that Crowley, who drifted ashore at Baiting Hollow, approximately forty miles from the site of the disaster, had suffered severe frost bite and, according to the newspaper, was likely to lose his toes and one of his fingers.

Despite their precarious condition, Crowley and the other survivors of the *Lexington* were indeed fortunate. They at least lived to tell the tale of a disaster which scattered bodies, baggage, and wreckage over a wide area. Along fifteen miles of the North Shore, guards were placed on duty to prevent a recurrence of the plundering which had taken place during the first few days after the *Lexington* disaster when baggage washed up on the beaches. A number of passengers on board were businessmen, and some were carrying large amounts of money.

The Heyday

Theft and various other forms of unlawful behavior were no strangers to the Sound steamers. By the late nineteenth century the steamboats had become the subways of the Sound, replete with a full complement of pickpockets, cardsharps, and prostitutes. On the large steamboats of the late 1800s, lawbreakers could easily become lost in the crowd. There was so much activity that people paid little attention to one another. Instead, they may have been gazing with wonder at the electric lights if they were sailing on the *Pilgrim* in the late 1880s. At the very least, they were concentrating on the elaborate interiors of the boats, particularly if they were traveling on one of the vessels of the Narragansett Steamship Company, a venture launched by Jay Gould and Jim Fiske in the 1860s. With their uniformed personnel and beautiful salons, they had offered stiff competition to the older Fall River Line engaged since 1847 in carrying passengers up the Sound from New York to New England. By 1879, as a result of a price war, the fare from New York to Boston fell to one dollar. The low price proved so attractive that passengers used to taking the railroad switched to the Sound steamers.

Throughout the nineteenth century the steamboats remained a viable alternative to land travel for freight as well as for passengers. Some of the more ambitious freight-carrying ventures fell through, such as the Farmington Canal Company's plan to build a canal from New Haven to Southwick, Massachusetts, where it would join the Hamden and Hampshire Canal and eventually link up with the St. Lawrence River; but the Sound remained an important highway for cargo. Even the unsuccessful venture of the Farmington Canal had positive side effects. When the part of the canal closest to the Sound was completed in 1828, New Haven's economy received a badly needed boost. Like many Sound shore communities on the mainland, New Haven had suffered during the War of 1812, and the canal appeared to hold the key to economic recovery. Thus, when the first boat went through to the Sound, people gathered all along the canal lighting bonfires to signify their approval of the venture. Although the venture was unsuccessful, the spectacle of people assembling to greet a canal boat symbolized the idea of Long Island Sound as a mammoth economic artery pumping life into the communities along its shores. Later in the century steamboats accomplished what the Farmington Canal could not achieve. They tied the Sound shore communities to one another and to New York City in a new and binding relationship unlike anything which had existed in the seventeenth and eighteenth centuries.

The Coming of the Railroad

Another aspect of the new relationship between the Sound shore communities and the great metropolis beyond Hell Gate was the railroad. Being faster and, in winter when the Sound was choked with ice, safer and more pleasant, this new iron monster would become the rival of the graceful steamboats plying the waters. Furthermore, early railroad men like the directors of the newly founded Long Island Rail Road* Company dreamed of a new connection between New York City and Boston, to be superior to existing routes. They were planning a line to be built across the length of Long Island from Brooklyn to Greenport, at that time a small fishing village with a good, well-sheltered harbor. Here, transferring to a steamboat, passengers would cross over to Stonington, whence a train would take them to Boston. Proper arrangements would be made to coordinate the operations of boat and train to insure a swift and comfortable trip all the way from New York to Boston.

Origin of the Long Island Rail Road

Of course, much discussion of detail preceded this ambitious undertaking. Would it justify the hopes of high financial return expected by the investors? Much depended on how the public would react to being ferried across the Sound. The chief engineer of the railroad, in a report dealing with the prospects of the undertaking, pointed out that the crossing from Greenport to Stonington would take two hours, and if proper accommodations were provided for the passengers, the service could be very lucrative. As for other aspects of the financial picture, the report described the solid foundations of the venture and the excellent circumstances of its inception. New York and Brooklyn, it said, were the mainsprings of energy and enterprise,

---

*Only recently was the name officially changed to the conventional one word form.

43

with a population of over 360,000. The railroad would cross Long Island, inhabited by 100,000. Over its entire length, a distance of more than one hundred miles, there would not be a single bridge and the grades would never exceed ten feet per mile.

The calculations regarding the time the trip would take envisaged a run from end to end in 5 hours; the boat would cross the Sound in 2 hours, and the rail trip from Stonington to Boston would require 4½ hours, total of 11½ hours. This was a vast improvement over the all-water route. Indeed, even the speediest steamboat took almost twice as long to cover the distance between New York and Boston by way of Stonington. In July 1827 the *Connecticut* had made the run to Boston in 23 hours. This was such a triumph that the newspapers not only took note of it but also editorialized on the American penchant for speed. Imagine, then, how the thought of traveling from New York to Boston in only 11½ hours must have captivated the public's mind.

One important factor in making the New York–Boston run in just under twelve hours feasible was the absence of steep grades and bridges on the proposed railroad from Brooklyn to Greenport. From the engineering viewpoint, this advantage offered the clue to the understanding of the whole concept. The originators of this project had one aim in view: to reach the point on the seashore closest to the railroad head across the Sound. In Boston, the promoters of the project pursued the same aim, so that for them too the Sound was the chief attraction. By crossing it, both could reach their objective in the shortest possible time: New York City and Boston respectively.

However, they overlooked one thing. The hundred thousand inhabitants of Long Island, referred to in the chief engineer's report, were not considered. Their interests were not going to be served. The railroad's directors deliberately selected a stretch of land almost totally uninhabited at the time. The right of way was to run parallel to the South Shore, about six to seven miles to the north of it, following the foot of the central ridge formed by the terminal moraine of the glacier which in geological times had covered the island. Its edge ran in an east-west direction. The debris of the melt, piling up during the centuries of this glacial epoch eventually formed the backbone of Long Island. The run-off of this process left behind sand and gravel mixed with clay, forming a poor soil now covered with pine and oak of stunted growth, known to the inhabitants as the "barrens." It could sustain no agriculture, only a few livestock. It produced abundant firewood, however, an item strongly influencing the calculations of the managers of the railroad, since at that time locomotives were fired exclusively with wood. Not suitable for agriculture, land on this featureless, flat plain could be acquired at nominal cost and, moreover, offered ideal terrain for rapid construction and fast operation, the point being to reach the Sound and not to serve the local area or promote its interest.

Enthusiasm generated by these seemingly excellent features explains the hasty

44

promotion of the Long Island Rail Road. Pursuing distant aims which had little to do with the ground on which rails were laid, the enterprise in the long run was bound to face serious consequences. It was a bridge spanning a void rather than an economic artery filling the void with life. One of the local inhabitants traveling on the Long Island Rail Road soon after its opening in 1844 made the following observation: "After leaving Jamaica, you scarcely see a village or a farm of good land, till you reach the terminus; but barren plains or forests of scrub oak, or stunted pine, environ the traveler on either hand."[5] At that time the population of Long Island east of Brooklyn concentrated on the seacoast, the only region where transportation, by boat, was available. No one cared to explore the roadless scrub-covered interior but a few woodcutters, hunters, or trappers. In the eastern parts of the island, far from the urban centers of the west, the people lived in complete isolation amid primitive conditions unaffected by much change since colonial times. They were not abreast of events, nor did they care to be. Thus, when the railroad came, they refused to believe what they saw until, as one observer relates:[6]

They beheld with their own eyes, the coumbrous train of cars, drawn by an iron horse, spouting forth smoke and steam, passing like a steed of lightning through their forests and fields, with such velocity that they could not tell whether the countenances of the passengers were human, celestial or infernal, they would not believe that a Rail Road had power almost to annihilate time and space.

The Threat of Competition

Now that the railroad had cut lengthwise through Long Island in a straight line almost as if drawn by a ruler, the big question arising in the minds of its directors was: What if another ambitious promoter should decide to connect New York with Stonington by a continuous, direct rail line linking up with the railroad to Boston? If this came about, and the possibility must have been obvious to the directors of the Long Island Rail Road, their own line would be doomed. It would be impossible to compete against a swift all-rail route. However, they simply ruled out such a possibility. Wishful thinking blinded them to reality.

The outwash of the great ice sheet on Long Island left behind a plain ideal for railroad construction. When during its many wanderings back and forth the glacier stopped on the line forming the coast of Connecticut and Westchester County, it left behind deep inlets and gullies, ridges alternating with rivers ending in swampy estuaries, all serious obstacles to profitable railroad construction. Many fills and cuts would be required, with steep grades and sharp curves adding to the cost and reducing the speed of the trains. As it appeared to the planners of the Long Island Rail Road when they began their project in 1834, building a railroad on the mainland side of the Sound would be a foolhardy venture. At any rate, by the time such a direct connection between New York and Boston could be successfully completed, the Long Island Rail Road would have made enough profits.

Premature Celebrations

Work went ahead vigorously as soon as the money was available. The line had reached Hicksville when the panic of 1837 caused a sudden disruption of all business activity. Funds dried up and construction came to a halt. It was four years before new financing made it possible to resume work. Pushing ahead again as hard as possible, the builders hired large numbers of workers, Irish immigrants as well as local laborers. They were well fed. An eyewitness observed that the workmen had unpeeled boiled potatoes for breakfast, with coffee, and the same for dinner plus corned beef and sometimes bread. Groups of them sat at tables loaded with huge dishpans full of food.

After long delays and hard work, the completion of the line called for great celebrations. On July 27 1844 a special three-section train arrived at Greenport with the president of the railroad, George B. Fisk, on board as the host of the festivities. Many guests were invited from the vicinity, some of them coming from as far as Montauk. Others came from New York, while additional visitors boarded the train en route to Greenport. As one of the passengers, a newspaper reporter, observed, brokers, editors, directors, and stockholders crowded the train, which at times traveled at forty-five miles per hour, making the entire trip in less than four hours. A local weekly reported that a special announcement assured the visitors that the company expected two large steamers to be ready the following week to take passengers across the Sound.

To accommodate the guests, most of them supposed to be prospective customers, the management set up a large tent. Under it were tables laden with provisions brought from New York including forty baskets of champagne and half a case of brandy. After the festivities were over, many of the guests could not get up, let alone walk, and had to be carried to the railroad cars. Such celebrations were commonplace in the early railroad age after the completion of an important line. People were simply carried away by the unbelievable reduction of time it took to go from one place to the other. Before the coming of the railroad, the absence of passable roads had made travel over even comparatively short distances a major expedition. Now, as if by magic, it took just a few hours.

In the celebrations at Greenport there must have been, aside from great elation coming from a sense of accomplishment, a feeling of triumph over the rivals operating steamboats in the Sound. The directors of the Long Island Rail Road must have been confident about the future of the great undertaking they headed. On August 10 1844 the company announced in the papers that the railroad had purchased the steamboats *New Haven, Worcester,* and *Cleopatra* from Mr. Vanderbilt. They were quickly put into the cross-Sound service.

The great plan of the Long Island Rail Road finally became reality. Ten years had passed since the ground-breaking ceremony when the directors turned the first shovelful of earth in Jamaica at the start of construction. But ambitious promoters completed many projects in New England during this decade and made plans to complete many more. The direct all-rail route between New York and Boston occupied a prominent place among these projects, but the difficulties in the way of realizing this old dream were still too great. On Long Island there was no need to worry about the immediate threat of competition coming from across the Sound, or so it seemed, at any rate.

Yet on the other side of the Sound mainlanders were wondering about the viability of the Long Island Rail Road's new route to Boston. The *New Haven Daily Herald* announced the opening of the cross-Sound service from Greenport by stating that the entire trip from New York to Boston was supposed to require ten to twelve hours, but the newspaper warned that the operators of the railroad "may be able to effect this at some seasons, but it is our opinion that one third of the year they will not be able to go it at all. The true and only effectual route will be through New Haven.—All experience teaches this."[7]

It seems that a little bit of jealousy clouded the interpretation of the *New Haven Daily Herald,* because it reported in a subsequent issue:[8]

The Long Island Railroad runs across so level and barren a country for the most part, that a greater speed may be safely attained than perhaps on any other railroad in the country. The principal drawback upon this route, so far as the long travel is concerned, is the necessity of crossing the Sound in the most dangerous part of it, where, with an easterly or southeasterly wind,—the only one which "kicks up" a heavy sea in the Sound,—it will strike the boats abeam and expose them to more inconvenience and danger than the whole trip from New York to Norwich or Stonington, with the wind ahead.

Although the *New Haven Daily Herald* was reflecting the bias of many New Haven citizens who imagined their city to be a great hub of transportation on the Sound, it was not an altogether inaccurate assessment. The route which the Long Island Rail Road Company's steamboats would use in crossing the Sound brought them beyond the sheltering arm of the North Fork of Long Island and out into the Race, that dangerous tidal turbulence in the vicinity of Block Island. The *New Haven Daily Herald's* criticism of the Long Island Rail Road Company's route to Boston was well founded, but this does not preclude the fact that it was motivated by the intense rivalry developing between communities on opposite sides of the Sound as they struggled to gain a monopoly of the transit revenues whether from steamboat or railroads.

In spite of the obvious portents, the directors of the Long Island Rail Road preferred to ignore the threat of a through line being built to Boston. They went on painting a rosy picture of the future in their annual report to the stockholders. Great hopes were attached particularly to the importance of Newport, Rhode Island, a rapidly growing resort town, and Fall River, Massachusetts, a great textile manu-facturing center. The 1845 annual report to the stockholders pointed out that the Long Island Rail Road would bring this whole area which offered so much promise within seven and a half hours of New York. While admitting the existence of great competition to all these points, the report hopefully expected an increase of revenues as soon as the railroad tied these cities into its expanding sphere of connections. The same steamer which would connect Newport and Fall River could, through the building of a new railroad, open up communications with Cape Cod and Plymouth, Massachusetts. Norwich, Connecticut, could also be included among the points of permanent contact. Strangely but consistently all these optimistic prognostications were based on the assumption that no all-rail route could be built between these New England points and New York City.

As for local Long Island interests, which heretofore had been completely neglected, the report mentioned that an agreement had been reached with the Post Office Department whereby stage routes running along the shore were to operate cross lines intersecting the railroad at various points. Great expectations were also attached to the development of Greenport as a future watering place rivaling Newport. On still another point, the time factor, the report was equally optimistic. Indeed, it stated that the running time of the rains exceeded all expectations. With an average of three hours and forty minutes, they were accomplishing "more in speed than had previously been effected on the continent of America."[9] As for the ferry service across the Sound, the performance left nothing to be desired: "The crossings have been made in all weather with the utmost certainty."[10]

The Impossible Becomes Reality

Such optimistic evaluations of the Long Island Rail Road's cross-Sound service would give way to extreme pessimism within three years when the New Haven Railroad Company opened a through line on the mainland side of the Sound. This was the railroad which the Long Island's directors said could never be built because of the difficulty of bridging the rivers flowing into the Sound. Engineering problems notwithstanding, the "impossible" became a fact, ruining the Long Island Rail Road's Boston service and reducing the railroad itself to a local line. The rosy expectations of the mid-thirties and the jolly celebrations at the opening of the through service gave way to dejection over lack of traffic and falling revenues. The "barrens" could not support a line designed to offer luxury service of express trains between two of the foremost cities of the nation. In 1850 the Long Island Rail Road went bankrupt and had to be reorganized. But this was only the beginning of an endless series of woes besetting the railroad which was so ill-conceived as to base its prosperity on the fickle waves of the Sound.

48

For this waterway, despite all the advances in transportation altering its economic importance, remained the often stormy and sometimes treacherous sea. Again and again she reminded man in the age of steam that his proud and powerful boats could, in minutes, turn into fragile shells. As we read in the poem "The Loss of the *Lexington*"[11]

> The Shrieking mother clasped the shivering child—
> The pallid maid her flowing ringlets tore:
> The tortured youth yelled startingly and wild;
>   And men bowed down that never prayed before!
> No help was nigh in that dread hour of gloom—
>   No life for them, nor hope beyond, save heaven.

Steam may have mastered the Sound in the nineteenth century, but in many ways the "American Mediterranean" was still very much its own master.

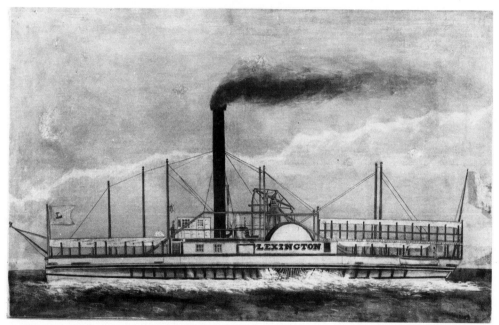

The ill-fated Lexington (Mariner's Museum, Newport News, Va.)

The burning of the Seawanhaka, 1880 (Harper's Weekly)

4

Railroads and Resorts on the Sound
1848-1900

BY THE MID-NINETEENTH CENTURY STEAM POWER HAD HARNESSED LONG ISLAND SOUND, its waters and its shores. The Sound's ferocious waves still challenged the steamboats, but new developments appeared on the shores, creating rapid changes in the overall transportation picture.

During the thirties and early forties the people of Connecticut realized that their state had been by-passed as two rapid transportation lines were opened between New York City and Boston, combining steamboat and railroad service. One connected Boston via railroad with Providence, Rhode Island, from there continuing by fast steamboats to New York. The other, as we have seen in the previous chapter, operated with two rail lines: one between Boston and Stonington, the other between New York and Greenport. Steamboats provided the link across the Sound. Neither of these connections touched the state of Connecticut except at its most easterly corner, at Stonington. The state itself, wedged in between its larger and more populous neighbors, faced the possibility of being completely relegated to the backwater of this rapidly developing transportation revolution.

New Railroad Developments

The obvious thing to do was to build an all-rail line across the state with one terminal at New York and the other in Boston. But the great vision such an undertaking required seems to have been lacking at this time in Connecticut. The plans to build railroads, if there were any, did not encompass great distances. Altogether, the beginnings of railroading in Connecticut were slow, the aims uncertain, and

51

construction piecemeal. Yet when the trans-state railroad linking New York and Boston finally became reality in 1848 by connecting New Haven with New York, the advantages of the new mode of travel were quite properly appreciated. A newspaper account of one of the first runs from New York to New Haven in December 1848 pointed to the obvious benefits: "The convenience of this new route will be hailed as a great public benefit, particularly at this season of the year, when the navigation of the Sound is so uncertain and dangerous."[1] Another newsper commenting on the trip stated:[2]

The train consisted of five large cars, containing over 200 passengers. The trip was rendered extremely difficult from the damp snow upon the track. There was a delay at Bridgeport and also at Williams Bridge. . . . The running time was within four hours—very respectable considering the necessity for prudence and the extremely bad state of the road consequent upon the late severe storm.

Happily for the directors of the New Haven Railroad, there had been a big snowstorm a few days before, and the inclement weather provided an excellent setting in which to demonstrate the superiority of a trip by rail over a voyage on the ice-clogged Sound. The citizens of New Haven who read about the successful trip in their local newspapers were probably as thrilled by it as the directors of the railroad, because the train trip signified the completion of the arterial route between New York and Boston. It ran through the middle of the state, following the valley of the Connecticut River and then, turning west at New Haven, tapped and boosted the trade of the state's Sound shore communities. Now the state and, even more, its shoreline facing the Sound were in the mainstream of development.

History of the New Haven Railroad

Immediately after its opening the line was well patronized fully justifying the high expectations of its promoters. But building the line had been quite a difficult task, even though the need for it was obvious to everyone. After many preliminaries a short piece, although destined later to become part of the main trans-Connecticut line, was completed in 1838. It ran between New Haven and Meriden, a distance of eighteen miles. As the very first railroad in the state, it functioned at that time only as a feeder to the steamboats operating from New Haven. Even with such a small and subordinate role, the line aroused bitter opposition. Hartford had direct steamboat connections with New York by way of the Connecticut River and the Sound, and the city did not want to lose out to New Haven, especially in the winter when the river would be closed to ships. The Farmington Canal and Turnpike interests were also inimical to the railroad project, but the state, realizing its importance, granted the charter for the railroad in 1833.

52

Obviously, New Haven would be the principal beneficiary of the railroad, since it would be its terminal point on the Sound. Extending the line farther north would also benefit Hartford by providing quick year-round service to the port of New Haven. After overcoming initial opposition, the two cities were linked in 1839. Soon after, the line's directors began to consider extension to Springfield, a station on the great arterial railroad between Boston and Albany, creating through this junction an all-rail connection between New Haven and Boston. When this link was completed in 1844, the question naturally arose, why not establish a railroad along the Sound to link New York City with New Haven? Initially the project created little interest, since steamboats offered excellent service and accommodations to New York. The Sound-side railroad scheme might therefore have been stillborn had it not been for the fact that Bridgeport, Stamford, and Norwalk were ready to challenge the Hartford and New Haven's role as a main line by building railroads into the interior along the Norwalk, Housatonic, and Naugatuck rivers. Being closer to New York than was New Haven, these cities could easily capture parts of the New York traffic and be serious competitors to New Haven. Also, if the proposed railroads in the Housatonic and Naugatuck valleys were to be extended far enough, they could offer an alternate rail route to Boston, even to Albany during the winter months when the Hudson River steamboats were impeded by ice.

Apprehensive of the possibility of New York traffic being diverted to new railroads in the interior of the state, the New Haven interests decided to build a railroad along the Sound from their city to New York. In this way New Haven would become the focal point of rail transportation, and the railroads running through the interior of the state would become merely feeders to the shore line. With this aim in mind the Connecticut legislature granted a charter for the New York and New Haven Railroad in 1844. As expected, the difficulties of the terrain made the building of the new line quite expensive. The estimates were overrun by a wide margin, and the railroad's entrance to New York City was opposed by the Westchester Turnpike Company and the Harlem River Railroad. In order to overcome this opposition and to obtain the necessary charter from the New York State legislature, the New Haven Railroad had to purchase the consent of the two New York companies.

Early Difficulties

Financial problems continued to plague the New Haven Railroad even after its through service along the Sound opened in 1848. Some of these difficulties arose from a dreadful accident which occurred in May 1853 when a train crowded with doctors returning from a medical convention in New York plunged through an open drawbridge at Norwalk. Following the accident, the *New Haven Register* reported:[3]

Many of the citizens of Norwalk witnessed the approach of the train, and some of them called out to the engineer to stop, as they foresaw the danger; but no heed was taken, and from the speed at which the locomotive was then going, it was impossible to stop so to avoid the accident.

Additional details were provided by the *Norwalk Gazette:*[4]

The draw had been raised to let the steamer *Pacific* pass through on her way to New York. The locomotive, baggage car and two passenger cars were precipitated into the river below, a distance of some twenty feet, and all the passengers buried beneath water, nearly all of whom were instantly killed or drowned, before assistance could reach them.

According to the New Haven *Daily Palladium,* "The cars were literally torn to atoms, and piles of seats and pieces of wood are now to be seen reared up like monuments."[5] The *Daily Palladium* also noted that[6]

Preparations were made this morning to drag the Creek, which is very deep, for the discovery of the bodies of the missing, and cannons were fired at intervals for the purpose of disturbing the waters and compelling them to yield up the dead.

Court proceedings following this terrible accident revealed that the engineer and conductor of the train had simply disregarded or overlooked plainly visible signals. As a result, the New Haven Railroad had to pay damages totaling a quarter of a million dollars. The accident at Norwalk prevented the railroad from paying dividends for a time. One also wonders whether the New Haven temporarily raised prices to offset its financial difficulties, because an actual commutation ticket from Mamaroneck to New York found in an old scrapbook gives the price of thirty-six dollars for round-trip travel between December 30 1854 and June 30 1855, while another commutation ticket for the Mamaroneck–to–New York run for the period from June 30 1856 to December 31 1856 cost only twenty-five dollars. To compound the financial calamity, the railroad's president, Robert Schuyler, was involved in a stock-watering scheme and subsequently fled to Europe with over a million dollars in railroad funds. Things were so bad that "in the early days a stockholder . . . was never willing to admit that he owned a share. Owing to what were termed the Schuyler frauds and also to great losses occasioned by the Norwalk disaster, . . . stock had very little value."[7]

The Goal Achieved

Robert Schuyler's financial mismanagement notwithstanding, he succeeded, before absconding to Europe, in coming to terms with the Hartford and New Haven Railroad for the merger of that line with the New Haven in 1872. By that time the New Haven Railroad had also acquired a major influence in the Housatonic and Naugatuck railroads. This eliminated competition from the state's interior lines.

Along the Sound east of New Haven, the need to ferry passengers across both the Connecticut and Thames rivers constituted formidable obstacles to fast transportation. Part of the problem was overcome in the late 1880s when the New Haven leased the Shore Line Railroad which had recently completed a bridge over the Connecticut River. With this transaction the New Haven pushed its line as far as Providence, there to be barred from reaching Boston, the most eagerly coveted objective. Not until 1893, when the New Haven acquired the Old Colony, initially a local line bringing vacationers to Newport and Cape Cod but later including the Boston and Providence Railroad, could entry into the Bay State metropolis be realized. Before this success the New Haven, following its merger with the Hartford and New Haven, became the New York, New Haven and Hartford Railroad. All this took some hard bargaining and shrewd maneuvering.

Thereafter the New York, New Haven and Hartford dominated the scene on the mainland side of the Sound. Its main line running next to Long Island was double-tracked and then double-tracked again before the end of the century, to form a massive, sturdy stem from which branched feeder lines up into the great industrial areas of inland Connecticut. Although steamboats still carried considerable freight and those passengers who had the time to enjoy a leisurely and luxurious voyage on the Sound, the New Haven became, in time, one of the major carriers of the nation, and along its tracks a whole string of towns grew up, some of them inhabited mainly by commuters while others became centers of industry and commerce.

Long Island

On the other side of the Sound, it was a different story. There the Long Island Rail Road started out as a long steel arm reaching out across the Sound to Boston, but with the completion of the New Haven Railroad's line to New York and the linking of that line to Hartford, Springfield, and Boston, the Long Island Rail Road fell upon hard times. In 1850 it went into receivership. This seemed to be inevitable in view of the fact that the Long Island was now a route from Brooklyn to nowhere. The wastelands stretching from Jamaica and Hicksville to the east could not support a railroad, and it would be decades before the interior of the island would be settled. This being the case, the only way for the Long Island to extricate itself from its financial difficulties was to expand in the direction of the more settled western portions of the island.

In 1854 a branch line built to Syosset immediately attracted freight and passengers from the surrounding country. The railroad might have been on the upswing save that the new president, Oliver Charlick, lacked the personality to establish harmonious relations with the communities lying in the direction of expansion. For example, he could not come to any agreement with the towns on the North Shore regarding the construction of a line along the Sound. Admittedly, the construction of such a line posed problems. The terrain was difficult, with many deep inlets from the Sound separated by high ridges. Deep cuts, high fills, and steep grades would be required in this area. The Long Island was not willing to undertake such a complex project until it was driven against the wall by a competitor, the North Shore Railroad, which by 1868 reached Glen Cove.

Realizing the seriousness of this development, Charlick quickly began extending the Syosset branch. Planning to reach Northport, he forced the North Shore Railroad to drop its original plan to expand eastward and instead to build in a northeasterly direction to Port Washington. Charlick also managed to get the citizens of Smithtown to part with fifty thousand dollars as his condition for extending the Long Island to that area and farther east to Port Jefferson, a thriving Sound-side city with shipping lines and a shipbuilding industry.

The Smithtown maneuver created much ill will for the Long Island Rail Road, but probably the worst scheme of all involved a long line cutting diagonally across the island from Flushing to Babylon. It competed with all other lines, creating chaos. Such contests explain "very much that is hard to understand in the tangled web of corporations, railroad tracks and abandoned tracks of the Long Island."[8] Although Charlick was subsequently removed, the Long Island Rail Road went through a series of receiverships. By that time[9]

a network of lines crossed and intersected each other at numerous points and competed on the same ground for travel and business that one railroad could easily handle. The effect of this situation was the sharpest kind of competition to secure the business with the result that wherever competition could reach it was done at ruinous rates.

Conrad Poppenhausen, Charlick's successor tried unsuccessfully to unify all the railroads of Long Island into one system. His successor, Austin Corbin, managed to create order out of a very confusing situation. He extended the North Shore line to Wading River in 1892, and the following year the main line reached Fort Pond Bay near Montauk. In 1895 the system was practically completed. Finally, in 1899, the New York State legislature granted the Long Island the right to build a tunnel under the East River. Purchased the following year by the Pennsylvania Railroad, the Long Island's network became integrated into the mainland transportation system. For all practical purposes Long Island ceased to be an island and became part of the economy of the New York metropolitan area.

Rise of the Resorts

Parallel with the development of its railroad came another important transformation of Long Island, particularly the North Shore. This area bordering the Sound now became a fashionable vacation spot. Although many of the vacationers preferred to travel to their destination by steamboat, the Long Island Rail Road played an increasing role in offering quick transportation to the beaches. Special pamphlets issued by the railroad described for instance the attractions of Sea Cliff, a locale[10]

appropriately named for its situation. The ground shelves down abruptly from the plateau to the waters of the Sound and the houses and streets rise, terrace on terrace, until they reach and crown the top of the bluff. Around this ridge the fresh breeze of the Sound is in constant play.

Sea Cliff apparently had so much to offer that the *Glen Cove Gazette* reported in 1873: "The attractions of bathing, sailing and fishing are of the sort to draw . . . people from brick and mortardom."[11] Farther to the east on Plum Island, the tourist found a remarkable mixture of agriculture and fishing. Plum Island is the visible part of a submerged ridge left behind as a terminal moraine by the ice sheet covering the major part of Long Island millions of years ago. Other portions of the submerged ridge jutting out of the water are Great Gull, Little Gull, and Fishers islands; these form a chain that runs in a northeasterly direction, joining the mainland at Watch Hill, Rhode Island. Like an irregular dotted line on the map, they demarcate the eastern end of Long Island Sound. Between them the tides rush back and forth, creating swift and dangerous currents and swirling eddies, long the menace of sailors. To help create a safe passage through Plum Gut, the channel separating Plum Island from Orient Point, the government erected a lighthouse in 1827.

In 1810 an enterprising man by the name of R. M. Jerome, noting the abundant grass, purchased the island and grew rich by raising thousands of sheep. Toward the middle of the nineteenth century, the island became "a Mecca of lovers of piscatorial sports,"[12] and since it had no flies or mosquitoes people began to arrive just for the sake of enjoying the peace and quiet of the bucolic surroundings. The lighthouse keeper kept boarders who, because of crowded conditions, often had to be accommodated in the boathouse or on fresh hay in the barn. The lighthouse burned sperm oil in a lamp outfitted with a burnished copper reflector. The shipwrecks, many of them coal barges, furnished wood and coal as fuel. One of the attractions for vacationers was:[13]

Sitting on the settee on the bluff by the flag pole, on a moon lit evening, watching the rushing whirling water, as it glows and glimmers in its phosphorescent light; the palatial steamers as they pass near the island all aglow with electric lights; the passing ship, like a phantom, goes sailing by, and you wonder what name she bears, and where bound. All is repose.

The recreational facilities throughout Long Island were so extensive that in 1880 one vacationer was prompted to write:[14]

A god-send to tired New Yorkers is old Long Island, with . . . incalculable stores of pleasure in bathing, sailing. . . . With big hotels and quiet farm houses, her lakes and bays . . . her rich cream and musical mosquitoes, if there is any more charming place for the idler in summer than Long Island, it certainly is not to be found anywhere near New York. . . . The Long Island Railroad has spread its arms all over the island, taking in scores of shady villages and settlements, and affording quick and comfortable transit to thousands of pleasant summer homes.

Eleven years after these words were written, the Long Island Rail Road inaugurated a ferry service across the Sound from Oyster Bay to Wilson's Cove in Norwalk, Connecticut. The new service utilized a specially constructed steel steamboat to transport not merely the passengers but the railroad cars themselves. Once on the mainland, the Long Island Rail Road cars were connected to the line of one of the New Haven Railroad's competitors for the trip to Boston. This unsuccessful attempt by Austin Corbin to recoup some of the line's losses lasted only ten months. Toward the end of the experiment, it was rumored that cardboard dummies were placed in the windows of the railroad cars as a face-saving technique. The dummies notwithstanding, the railroad ferry service from the North Shore to Connecticut failed to take hold, just as the Long Island Rail Road's scheme for creating a port of free entry at Montauk failed. There was at least something of a precedent for this latter concept, since another Long Island town, Sag Harbor, had been an official port of entry for the United States. By the mid-nineteenth century Sag Harbor had another distinction as well. It had become something of a tourist attraction for Connecticut residents who crossed the Sound on excursion steamers for a one-day trip to this charming whaling town. Excursion boats also provided one-day round-trip service from New Haven to New London and Norwich.

Farther south along the Connecticut coast, the Thimble Islands off Branford were attracting numerous visitors, including some who would not ordinarily have been able to afford a few weeks at a Sound-side hotel. The Madison Avenue Presbyterian Church of New York City leased Pot Island in the Thimble Islands group and brought working women and their children as well as single working girls to the hotel on that island for two-week vacations. The spirit of charity caught on, with the result that the mayor of Waterbury, Connecticut, who had a summer residence on one of the other islands, opened his home to the working girls of his city.

Poor working girls were not the only ones to enjoy the pleasures of the Sound. Farther down the coast of Connecticut, William Marcy Tweed, boss of the Tammany Hall Democratic machine in New York City, was savoring the delights of Round Island off Greenwich. "Boss" Tweed had learned of the island from several of his associates who, one summer's day in the 1860s, rented a sailboat at City Island for a brief excursion. The trip was extended to an overnight adventure when a storm blew the vessel off course. Landing at Round Island, the occupants of the rented sailboat decided to spend the night there. When they awoke the next morning to see the beauty of their surroundings, they reported that fact to "Boss" Tweed, who subsequently went out to Greenwich on the New Haven Railroad to inspect the island. Tweed liked the place so much that he leased Round Island as a campsite and later built a headquarters there for the Americus Club, which he and his associates established.

By the late 1860s Tweed was also involved in the steamboat business on Long Island Sound. His Greenwich and Rye Steamboat Company, established in 1866, ran boats to New York City. One of Tweed's steamboats, the *John Romer,* had the unusual distinction of overtaking the boat regarded as the fastest on the Sound, the *Seawanhaka,* serving the two North Shore communities of Roslyn and Sea Cliff with daily runs to Manhattan. On June 2 1867 the two ships raced each other down the Sound, and it must have been quite a race because one of the officers of the *John Romer* reported:[15]

The passengers and crew of both boats were . . . in a fever heat of excitement. . . . I think I never saw such a crazy lot as yelled at each other across the span of a dozen feet between the two boats. Women shook their parasols in the air and squealed like a flock of geese.

Previously the *John Romer* had always been overtaken by the *Seawanhaka* whenever the two boats met en route to New York, but now she surprised everyone by winning the race. After that memorable June day in 1867 she was never again bothered by the *Seawanhaka.*

More than a dozen years later, in 1880, fate dealt the *Seawanhaka* another severe blow. Just as tragedy hit another Sound steamer, the *Narragansett,* on June 11th of that year, the *Seawanhaka* with some three hundred passengers aboard caught fire in the East River on June 28th. Among the passengers were William R. Grace, at one time mayor of New York City, and Charles A. Dana, editor of the *New York Sun.* Both gentlemen were saved, as were most of the passengers, who heeded Captain Charles Smith's warnings not to jump until he had beached the boat on a shoal between Ward's and Randall's islands. Captain Smith, who remained at the helm throughout the ordeal, died a year later from the effects of the disaster. Before his death the residents of the North Shore, many of whom had commuted regularly to New York aboard the *Seawanhaka,* presented him with a purse in appreciation of his selfless action on the day of the disaster.

Among the grateful contributors there were undoubtedly a number of well-to-do people who used to take the *Seawanhaka* to such fashionable North Shore spots as Glen Cove, whose Hotel Pavilion had stables and a bowling alley. Similar hotels were going up in other Sound shore communities. In 1873 when the Sea Cliff Hotel, which boasted water in every room and a beautiful view, was nearing completion, a new line of steamboats including the *Arrowsmith* made two trips a day to New York. The fare was thirty-five cents, and commuter books were available to those who

59

went back and forth to the city frequently. The *Glen Cove Gazette* reported in the same year, 1873, that "Garvies Point and Sea Cliff are the fashionable resorts for bathers."[16]

The Era of Mansion Building

On the mainland side of the Sound, steamboats were bringing visitors to the famed resort of Glen Island in New Rochelle. With its magnificent floral displays and beautiful tropical birds, Glen Island was a favorite spot for travelers from both sides of the Sound. Rye Beach was another big attraction along the Sound shore of Westchester County. Large hotels catered to summer visitors, sometimes to the dismay of Rye's other summer residents, the small but growing group of millionaires who, in the second half of the nineteenth century, built great estates along the Sound. Unlike some other Sound shore communities, Rye did not have to wait until the late nineteenth century for her mansions. Already in the 1830s Peter Augustus Jay, son of John Jay, one of America's founding fathers, built a magnificent home which is still standing on the Post Road. The property, extending down to the Sound, was originally purchased by Peter Augustus's paternal grandfather, the father of John Jay. A well-to-do businessman, he moved his family from New York City to Rye in the early 1700s to escape the further ravages of a yellow fever epidemic which had already blinded two of his children.

Most of the wealthy businessmen who fled to Rye and other Sound shore communities in the second half of the nineteenth century were not seeking a healthier atmosphere for their families, except perhaps for the feeling of well-being often induced by living near the sea. Many were undoubtedly attempting to display their wealth, at least to their fellow millionaires, while others probably wanted nothing more than a quiet retreat in the country where they could escape for the weekend. The dependable transportation provided by the New Haven and Long Island railroads made a weekend in the country feasible by the mid-nineteenth century. Following a fairly brief journey on the train, the tired executive would arrive at his destination assured of the fact that even if he left his Sound-side retreat on Monday morning he would be in time for a full weeks' work in New York. Train travel on the Long Island and the New Haven was really quite acceptable, although on the latter line, at least in the early days, certain amenities were lacking. For example:[17]

The club car was unknown . . . but certain commuters who desired to play cards occupied their own camp chairs in the baggage car. These chairs were in charge of the baggage master, who had little else to do and his compensation was a generous Christmas collection. This was the origin of the present club car service.

Even after full-fledged club cars were introduced, some people preferred to

60

avoid the railroad, and for a very good reason if we are to believe the lament of one harried New Haven passenger who wrote:[18]

I protest against passengers having to stand up for want of seats, at any time, but particularly when there are empty cars in the same train locked. This happened on the 3d August; on the evening of the same day, an old gentleman, past 70, an admiral in a foreign service, coming out to see me, had to stand up the whole distance, and not being able to find out when he got to New Rochelle, was carried to Mamaroneck; the next day he had to stand again; after dark, on his return.

This complaint was included in a letter to the president of the railroad along with complaints about the safety of the tracks and about "an army of dirty, ragged, shoeless boys"[19] who boarded the trains in New York to sell candy. The commuter who wrote this letter may have been contemplating something bigger than a one-man protest, because he took the trouble to have the letter printed in booklet form.

Commuting on the New Haven Railroad being somewhat less than perfect, some people who could afford to do so traveled back and forth to their estates by yacht. By the late nineteenth century yacht clubs were proliferating along the Sound. One of the first to be established was the Larchmont Yacht Club, founded in 1880. Catering to sailing enthusiasts, the club hosted an annual regatta and brought guests up from New York on its own steamer the *Crystal Stream,* which made the trip from 23rd Street in the East River to Larchmont in a little over two hours. Sailing yachts setting out from Larchmont were part of an advance wave of boats using the Sound primarily for recreation. In time steam yachts would join the procession. The American Yacht Club in Rye was the first steam yacht club on the Sound. It came into being in 1887 because one of its founders, Jay Gould, had been prevented from joining some of the other yacht clubs.

Discrimination among millionaires notwithstanding, it would appear that as early as the nineteenth century segments of the shoreline bordering the Sound were being set apart as the private preserve of the rich. In Mamaroneck, which boasts a fine harbor on Long Island Sound, E. H. Weatherbee built a mansion resembling the Elizabethan castles of his ancestral homeland. His estate, Wayte's Court, had a Japanese-style library and a dining room ceiling taken from a Spanish convent. And to enhance wealth with prestige, he had the music room made as an exact replica of the main drawing room in one of Queen Elizabeth's favorite castles.

Another Mamaroneck resident, the oil magnate Henry M. Flagler, personally supervised the design, construction, and interior decoration of his mansion at Orienta Point right down to the crystal chandeliers. After the workmen had executed any design for the interior of the house Flagler demanded that the molds be destroyed in order that no one else might possess the same fixture. Apparently Flagler had other maverick ideas. Afraid lest he die once his dream house was completed, he decided to challenge fate. Completing one section of the house, he waited two years before building the next part, hoping to prolong his life. According to reports, his seawall cost half a million dollars, and to create the proper kind of Sound-side beach at the estate he ordered the sand brought there from New Jersey. Flagler is said to have spent many hours in his mansion conferring with John D. Rockefeller, his partner in the oil business. He had power and influence and loved flattery. On those occasions when business compelled him to go to New York, he could always be sure of transportation. The directors of the New Haven Railroad ordered every train, including expresses, to stop at Mamaroneck for an instant to see if Mr. Flagler was on the platform.

At this time, the heyday of estate building, some millionaires strove to attain the ideal condition, that is, to combine their love for the Sound with the need to be close to the city. Among them we find the sugar magnate H. O. Havemeyer, who had a Sound-side residence adjoining Fort Schuyler. Another Sound-lover, Collis P. Huntington of Southern Pacific Railroad fame, had a thirty-acre estate in the same area. In the 1880s Mr. Huntington was spending "about seven months of every year at his charming country-seat at Throgg's Neck, on Long Island Sound, whence he can reach his business and return every day."[20] Typical of the wealthy Sound-side proprietors, its owner,[21]

securing the best talent and sparing neither time nor money, has continued to adorn and improve the house and lands until at present—with its system of water, its gasworks, its private wharf, at which large vessels are occasionally moored, its stables, conservatories, farm buildings, pastures, shady walks, gardens and flowers—it is a model residence and a place well fitted to divert the fancy, restore the strength and rest the heart. . . .

One of Mr. Huntington's contemporaries wrote:[22]

His country residence, at Throgg's Neck, is a refuge and great source of pleasure to him. From the broad verandah of the house a neatly-kept lawn slopes away under the branches of noble trees down to the water of the Sound, and here, on a clear day or a pleasant evening, Mr. Huntington . . . may often be seen strolling up and down in conversation with friends, or watching the steamboats and sailing-vessels as they pass. . . .

Some millionaires such as T. C. Benedict, whose estate The Maples bordered Long Island Sound in Greenwich, demonstrated their generosity by placing their yachts at the disposal of their friends.

But one of the most famous voyages made by Benedict's yacht *Oneida* was anything but a pleasure trip for the principal guest, President Grover Cleveland. During his second term he developed a malignant growth on the roof of his mouth. An operation proved necessary. With the country caught in the grip of a huge financial disaster, the panic of 1893, word of the president's impending surgery might cause the American economy to take another nose dive. So as not to alarm the public the operation was to be performed on the yacht *Oneida*. Fortunately, Long Island Sound was perfectly calm on the day selected for surgery, July 1 1893. Indeed, there seemed to be nothing unusual about the president taking a few days off to go sailing. Within two days of the operation, Cleveland was up and about. When on July 4th the *Oneida* made a stop at Sag Harbor, reporters were on hand to inquire about the nature of the president's trip. The official announcement that Mr. Cleveland had been treated for ulcerated teeth aboard the yacht seemed to satisfy the public's curiosity. Although minor rumors spread regarding the president's health, the *New York Times* commented two months after the operation: "The President seems a trifle thinner than he was a year ago, but on his face is a ruddy glow. His step . . . was elastic, his carriage erect, his actions bespoke a person enjoying perfect health."[23]

Thus ended the story of one president's famous cruise on the Sound, but it was not the end of presidential involvement with the inland sea. President Theodore Roosevelt maintained a summer White House at Oyster Bay, Long Island, where he and his family enjoyed the pleasures of boating and fishing. Built in the 1880s, the house was to have been called Leeholm for Roosevelt's first wife who died in childbirth on the same day his mother died. Following this double tragedy Roosevelt went West for a time, and after returning he renamed his estate Sagamore Hill. Here he spent summers with his family regretting the end of each season. Once he described the closing of the house in a humorous way in a letter to his son Kermit:[24]

Mother is as busy as possible putting up the house, and Ethel and I insist that she now eyes us both with a purely professional gaze, and secretly wishes she could wrap us up in a neatly pinned sheet with camphor balls inside.

Such good times enjoyed by Teddy Roosevelt and his family on the Sound were not limited to the rich. Anyone who could afford a train ticket or an excursion ticket on one of the steamboat lines could go to a Sound-side beach for a day, if not for a week or longer. One did not have to own an estate or frequent the famous resorts to enjoy the Sound, as a newspaper reporter noted when he stumbled upon an actors' club on High Island located in the Sound off New Rochelle. According to the reporter:[25]

The actors, men and women, visit the place on Sunday and hold high carnival in the way of pure, fresh air and all that sort of thing. There is no doubt about its being "high island" especially on Sunday.

Although one can legitimately question what the visitors to High Island were enjoying more, the Sound or the liquid refreshments they brought along on their Sunday picnics, there is no doubt that the Sound had become a recreational area during the second half of the nineteenth century. More and more yachts could be seen amidst the passenger- and freight-carrying steamboats, while on both shores there were beaches, resorts and the manicured estates of New York's elite. Thanks to the completion of the New Haven Railroad and the Long Island's North Shore

63

line, poor and rich alike could enjoy the pure water, the fresh air, the pleasant views, and all the other delights of a weekend or a vacation of a whole season. A new day had dawned for Daniel Webster's "Mediterranean," and the stage was set for even greater developments of an economic nature.

Theodore Roosevelt with Boy Scouts, Oyster Bay, 1918 (Nassau County Historical Museum)

5

Agriculture, Fishing, Whaling, and Industry

THROUGHOUT THE NINETEENTH CENTURY WHEN THE SOUND WAS BEING TRANS-
formed by steamships into a multilane highway linking New York with New
England, people who lived near the inland sea began to react to the new spirit of
progress symbolized by the introduction of steam. Hustle and bustle became commonplace at once-quiet ports. On the mainland side of the Sound, great cities
emerged as a result of the Industrial Revolution sweeping the country during the
nineteenth century. Elsewhere, however, the old ways persisted. On the North
Shore of Long Island, for example, fishing and agriculture were the mainstays of
the economy long after industry had become a way of life on the mainland.

Agriculture

The North Shore was well suited for agriculture. Its rich soil and long growing
season, a result of the benign influence of the sea, permitted farmers to raise
everything from wheat and potatoes to asparagus and fruit. Wheat growing in
New York State is supposed to have begun on Long Island. On the Hempstead
Plains wheat thrived. The sandy soil was good also for rye and hay. By 1840 the
island led the state in the production of the latter crop. Dairy products, too,
figured prominently in the island's agriculture. Here, as in other types of farming,
organization was the key to success. It was a similar case with sheep raising. The
island had one of the earliest woolen industries in New York. Somewhat later the
famous Long Island duckling, really an immigrant from mainland China, made its
debut and in time became a factor in the economy. Hogs were also important.

Many a Long Island pig was shipped to the West Indies in the nineteenth century. The popularity of the Long Island swine is explained by the fact that "on Long Island pigs grew fat on oysters, clams, and ocean refuse and literally overran the island."[1]

While the Long Island terrain was neither unique nor extraordinarily beautiful, it was more than adequate to sustain various types of cultivation as well as a diversified animal population. Some visitors such as Timothy Dwight, the peripatetic president of Yale College, making a trip to Long Island in the early nineteenth century, seemed to decry the absence of outstanding topographic features. He commented that there was "nothing bold and masculine"[2] about the North Shore and said further: "Long Island from Huntington to Southold and probably for a considerable distance further westward . . . is like the peninsula of Cape Cod . . . a vast body of yellow sand."[3] Dwight may not have considered the North Shore exciting or unusual, but he did note the sandy soil, a natural resource which encouraged production of huge quantities of hay and potatoes in such North Fork areas as Southold and Oyster Ponds. The Long Island historian Benjamin F. Thompson commented that in Oyster Ponds "the soil is not surpassed by any upon Long Island."[4] Nearby Southold also had soil known to be very favorable to agriculture. Potential farmers, especially New Englanders who came across the Sound to the North Fork, were pleasantly surprised to learn that the land was free from stones. Even in relatively poor areas like the Pine Barrens crossed by the Long Island Rail Road, agriculture was feasible. First proposed in 1860, the idea of developing this land failed to appeal to people until over four decades had passed. Then, in the early 1900s, the president of the Long Island Rail Road, Ralph Peters, sponsored the development of an experimental farm in the Pine Barrens. Without the use of chemical fertilizers[5]

three hundred and eighty varieties of plant growth were successfully developed or naturalized. This great number was experimented with in order to prove conclusively to the world at large the fact, well known to real Long Islanders, that any plant growable in the Temperate Zone could be developed far above the average in quality and further; many little known or entirely unknown growths of marked value in their native countries would readily naturalize with the particularly favorable conditions of Long Island climate and soil.

Farther westward on the island farmers were growing vegetables for the New York market. In Oyster Bay asparagus became an income-producing crop. Across the Sound in New Rochelle, blackberries were an important product. The "Lawton," or "New Rochelle," blackberry met with great success in the nineteenth century, as did the fruits grown at Prince's Linnean Botanic Gardens in Flushing. This was the first botanical garden in the United States, and the son of its founder excelled in the field of horticulture, writing a number of scientific works on fruits and flowers. The vast majority of Long Island farmers were not so advanced as the Princes, but they probably would have received a higher rating for diligence than their colleagues

66

across the Sound, about whom Horatio Gates Spafford said in the 1824 edition of his *Gazetteer of the State of New York:*[6]

It appears to me that the inhabitants along the eastern part of this County [Sound area of Westchester] are wanting in enterprise, if not in ingenuity, compared with their neighbors, of Connecticut, a remark particularly applicable to the use of water-power, in manufactures, and mill-work. While they make all sorts of notions and turn all their labor and water-power, and surplus agricultural products into those goods, and money, the Westchester men follow the plough, content, if the surplus of their farms enable them to become their purchasers.

Despite the situation in Westchester, at the very time Spafford recorded his comments farmers on both sides of the Sound were producing a wide variety of crops for both local consumption and the New York market. The relative ease with which farm products could be shipped to the great metropolis beyond Hell Gate explains the high level of production. Before the coming of the railroad, farmers had to take to the waterways. In fact, "Agricultural produce . . . in its first century and a half was chiefly sent to market over . . . the Sound. Land traffic—there was little of it—was by oxcart, horseback, and tedious travel on foot. . . ."[7] The advent of the railroad on Long Island and then on the mainland provided producers with an alternative, but well into the nineteenth century the Sound bore most of the traffic carrying crops to market.

Menhaden

The Sound helped the farmer in another way, too. It provided fertilizer for his fields. Seaweed cast up on both shores could be well used for this purpose. Fish would enrich the soil even more. "Mossbunkers," the local name for the menhaden, shad, or bony-fish spread regularly over the fields, would increase the yield spectacularly. Prior to the introduction of this method, Suffolk County could not raise enough grain even for local use. After adoption of the new idea, yields increased dramatically to the point where the county could export grain. Ezra L'Hommedieu, one of the leaders of the New York Agricultural Society and himself a Long Island farmer, wrote:[8]

In Suffolk County, some years ago, at Huntington, by manure, 52 bushels of wheat were raised by the acre. Since the practice in that county of manuring land by fish has been in use, between 40 to 50 bushels of wheat from one acre is not an uncommon crop. And by a late accidental experiment, it appears, that the product of grain from an acre will be in proportion to the quantity of this manure, and so far as to exceed any production we have heard of, in any part of the world.

L'Hommedieu went on to recount the story of a Riverhead farmer who used mossbunkers on his rye field. The results were so surprising that although a neighbor's sheep twice ate the grain right down to its roots the rye sprang back thicker than before.

With so many farmers using fish as organic manure, it appeared to some observers like L'Hommedieu that the supply would be depleted, but far from it! L'Hommedieu actually observed that the fish population increased, so that one draught often brought in a haul of 250,000. Another observer noted that on the eastern end of Long Island catching fish for fertilizer was[9]

regularly pursued, during a part of the summer, as ploughing and sowing, or mowing and reaping, in their appropriate seasons. For this purpose, the farmers of a neighbourhood form themselves into a company, sufficiently large to afford a relief of hands, every week, and having provided themselves with a large seine, boats, a fish-house on the shore, and every necessary convenience, the party on duty, take up their residence on the water-side; and it is impossible to convey to a stranger's mind, the immense product of a week's labour. A single haul of a seine has been calculated at 1,000,000 of fish.

The application of this huge haul of fish, primarily menhaden, to farms on the North Fork of Long Island greatly increased the value of the land. Furthermore, it demonstrated to both the original settlers who had come from New England and to newcomers that the Sound provided many benefits. The water not only yielded nourishment for human beings in the form of edible fish (and incidentally, many such species were hauled in with the menhaden or mossbunkers), but at the same time it provided enrichment for the soil.

Menhaden taken from the Sound helped seacoast residents in another way. They provided employment in the factories established to extract the oil used in tanning and dressing leather. The method was simple. When the menhaden were boiled in a big iron pot, the oil rose to the surface. By the mid-nineteenth century factories specializing in the production of oil and guano fertilizer from menhaden had sprung up in Greenport, on Shelter Island, in Groton, Connecticut, and on the Norwalk islands. Special boats were used to haul in the fish and then transport them to the factories on both sides of the Sound. In 1880 the total value of products made from the menhaden in the state of New York passed one million dollars, with much of the fish taken off the North Shore of Long Island. The guano manufactured in Southold, Riverhead, and Shelter Island rivaled in quality the famed Peruvian guano.

The menhaden industry had great disadvantages as well as advantages. The processing plants emitted a penetrating odor. People could live with the stench if they thought that the menhaden were aiding their fields. As an impartial observer noted: "The application of fish over extensive fields, as usually performed, taints the air to a great distance."[10] He added, however, that "it is not . . . a miasma which occasions sickness."[11] People also learned to endure the smell of the menhaden if they made their living in one of the fish-processing factories, but once the summer people inundated the North Shore of Long Island and the mainland side of the

Dissecting whales on the beach (Harper's Weekly)

The "Charles Morgan", last of the great whaling sailships, in a diorama at the Cold Spring Harbor Whaling Museum (Photo by Allan M. Eddy Jr.)

Summer scene at Lloyd Harbor, 1881 (Nassau County Historical Museum)

The steamship "New York", which carried Charles Dickens down the Sound (Samuel Ward Stanton, "American Steam Vessels")

Sound, factories moved out. The industry declined rapidly, facing the competition of chemical fertilizers and mineral oil.

The Lure of Whaling

The same fate befell the whaling industry, one of the oldest occupations of the people living around the Sound. When the first settlers arrived in the region of Long Island Sound, they found that the Indians had been practicing a primitive sort of whaling. Setting out in their dugout canoes, they would select females or their calves from the pack and poke them with their bone- or rock-tipped spears until the whale would tire. Then they would drive their quarry onto the beach where the squaws and children were waiting with knives to finish off the prize. For them practically every part of the huge body of the whale was an important source of their livelihood. They ate the meat, used the oil for cooking corn or peas, and fashioned utensils and weapons out of the bones or carved them into ornaments. The latter practice, called "'scrimshawing" by the white man, thus originated with the Indians.

The Indians never went any farther. The practice of going out onto the open sea to hunt whales remained unknown to them. They depended on whatever happened to come to their territory, whether it was a disabled animal drifting or one driven onto the beach by a severe storm. Circumstances forced the early settlers to maintain the same passive attitude and wait for drifting whales. They had better tools and implements than the Indians, but like the Indians they could not brave the open sea before they were able to build or buy seaworthy vessels. Until then they practiced whaling in Indian fashion, often employing experienced braves as harpooners or oarsmen in their alongshore ventures.

Unlike the Indians, the whites did not consider any part of the whale as good for food. It was the oil they were after, a precious substance in the absence of other suitable material to provide light in their homes. It could be sold or exchanged easily, and its universal value made it a sort of currency among the early colonists. Villagers living around the Sound are even said to have paid the salary of their ministers and teachers in whale oil. They obtained this precious substance from the blubber, a thick, fatty layer surrounding the whole body of the whale to insulate it from the cold and to give it buoyancy. This material, possessing a penetrating fishy smell, was rendered or "tried" in huge kettles. The work had to be done quickly, because the blubber of a dead animal lying on the beach would soon spoil. Cooking out the oil was a day and night affair in which the whole village participated. A fully grown whale could yield one hundred barrels, easily marketable in the Sound shore settlements or in Boston.

Some whales had a much more valuable substance, the so-called spermaceti oil, or sperm oil for short, located in the cavity of the head. Hardening into a waxy substance, it could be easily fashioned into candles. Sperm oil could also be used very effectively in softening leather. Being odorless, it had a value far exceeding that of ordinary blubber oil, and lucky were the captains and crew who caught sperm whales.

Nature provided a peculiar mechanism for this huge mammal of the sea to feed itself by. A whale could fill its mouth with fifty to sixty gallons of water, which was then forced out with its enormous tongue through a dense thicket formed by hundreds of "bones"—that is, long, flat, extremely elastic bony sticks. These "teeth" of the animal, though not used for chewing, screened out what was edible in the water. The bones found various uses in the manufacture of clothing especially women's corsets. Imbedded in the fabric as stays, they made the garment firm but elastic. Following the Civil War, when women's style called for a thin waistline, the bones became really valuable. A pound, formerly around twenty cents, would now fetch as much as six dollars. In fact, after the discovery of mineral oil in 1859 the whalebone supported the industry for decades. With the coming of rubber the bone lost its importance and followed the oil in its decline.

Knowing these precious substances to be contained within the huge body of the whale, men found the lure to hunt them irresistible. Yet it was a very dangerous profession. With one swing of its enormous tail, the frightened or enraged leviathan could break a boat or even a small ship and endanger the lives of the crew. Many an expedition, after sailing forth from the home port, was never seen or heard from again. Regular merchant vessels would follow a straight course, a "beaten path" as it were, and could help each other if necessary, but the whaler would wander around in search of his quarry. If he was shipwrecked, there was little hope of being rescued.

To counteract the loneliness of such solitary voyages, whalers, yielding to man's gregarious instincts, would congregate regularly at certain points of the globe as if by a silent understanding turning gradually to custom. One such point was the Hawaiian Islands, where the crews could rest, swap stories, and exchange information. As all fisherman know, Lady Luck frowns more often than she smiles, as we also learn from the letter of Captain Frederick A. Weld of Sag Harbor, commander of the ship *Italy*. Writing from Lahaina, Maui, on November 29 1845 to the ship's owner, David G. Floyd, he happily reported having fourteen hundred barrels of oil, including sperm oil, and ten thousand pounds of bone. He also mentioned that "some ships have not got a whale this season," adding that there were 190 ships there that season "and none has done remarkedly well."[12] With the American consul's certification, complete with seal and signature, the letter must have left Mr. Floyd satisfied with the report of his captain.

In spite of all the dangers, or perhaps because of them, the whaling industry attracted many men. Two types were particularly common among the crews:

young lads, often in their teens, who sought adventure and eventually careers as masters of ships, and the older generation, tough, rough, and mostly ignorant characters who had no family ties and no trade. Being rootless, they did not mind signing on for voyages which would sometimes take years. The rest of the crew consisted of skilled craftsmen, that is, carpenters, blacksmiths, coopers, and sailmakers. To preside over a motley crew of this kind required great skill and force of character. The captain, or master, was in charge in the full meaning of the word, and the success or failure of the voyage depended on his skill and judgment. His word had to be obeyed instantly, especially when maneuvering around a wounded and thrashing whale ready to crush its assailants. Whaling, in a way, was a great school and training ground for men of resourcefulness, sportsmanship, and a stout heart. It produced its own heroes with exploits of immense endurance and at times incredible daring. All this created a lore rich in legend woven into a brilliant web of fact and fiction, some of it dealing with the Sound and its daring sailors.

It must be remembered that the great packs of whales, migrating in the fall from the Arctic Sea southward to seek warmer waters, seldom ventured into Long Island Sound. Their habitat in the winter months was the open Atlantic. Prior to the American Revolution the South Shore of Long Island, Peconic Bay, Nantucket Island, and Rhode Island were in the forefront of development in the whaling industry. Sag Harbor on Long Island and New Bedford in Massachusetts distinguished themselves in this period and well into the middle of the nineteenth century, the heyday of whaling. By this time ports on Long Island were taking their share of the whaling business. With all its hazards, whaling could be a very lucrative business, and distance from the sea played no part in the ventures undertaken in increasing numbers in the ports of the Sound and in one North Shore locality in particular, Cold Spring Harbor. Though it lies about one hundred miles distant from the open sea, the businessmen of that town decided to organize a whaling company in 1839. The time required to cover the distance separating the town from the open sea had little importance for the whaler who was setting out on a voyage taking two to three years. For this reason, one should not be surprised to find in the register published by the U.S. Commission on Fish and Fisheries in 1878 that towns on the Hudson River so distant from the open sea as Newburgh, Poughkeepsie, or Hudson, N.Y., sent out whalers at regular intervals in the middle of the nineteenth century, the peak time of the industry.

Principal promoters of the company at Cold Spring Harbor were the great grandsons of Major Thomas Jones, a pioneer settler of what is today Massapequa. He obtained a license from the first governor of New York province in 1705 to take drift whales in an area including practically the whole of western Long Island. It was a virtual monopoly, for which he had to pay one-half of the catch to the governor. His descendants operated under better conditions. The company they founded in 1839 gradually acquired nine ships which sailed to all parts of the globe.

The best years for Cold Spring whaling were 1850–54, when the little town became one of the busiest shipping centers on the island. A cannon would be fired on one of the hills surrounding the harbor when a ship was approaching. Then the streets would come alive with people. Near the wharves there were shops and stores producing articles needed in the whaling trade around which the economic life of the town centered. On Bedlam Street, as the name suggests, were the inns catering to the sailors who contributed their share to the town's economy, but not always in a quiet and businesslike manner. On land captains and members of the crew mingled on an equal standing, a welcome relief from the harsh discipline reigning on board.

While Cold Spring Harbor occupied a unique position on the Long Island side of the Sound, the mainland side was dotted with whaling establishments. The register of the U.S. Commission on Fish and Fisheries lists a number of places sending out whalers on a regular basis. From east to west they were Stonington, Mystic, New London, Norwich, East Haddam, New Haven, and Bridgeport. Among them New London played an outstanding role. Whaling began early here but came to a complete halt during the wars with England, to revive again after 1814 together with the West Indian trade, which had developed on a larger scale before the Revolution. In contrast to the latter trade, marked there by "habits of dissipation, turbulence and reckless extravagance,"[13] the whaling industry encouraged "order, happiness and morality,"[14] reflecting its wholesome influence on the city. In 1854 whaling reached its culmination in New London. At that time the city ranked before Nantucket by more than one thousand tons, with only New Bedford exceeding New London in the trade. But overextension followed, and in 1847 the inevitable decline set in, reducing the number of ships from 71 to 66. The unemployed men responded to the siren call of the California gold rush with eagerness. In the years 1849–50 twenty-five whaling captains sailed around the Horn, taking with them many kindred spirits of the whaler in adventure and gambling. A similar situation existed in the other whaling towns of Long Island Sound, with the result that those people who passed up the chance to go to California and instead remained home turned to new economic pursuits, including the harvesting of oysters.

Oysters

Whales with their huge bodies and geyser-like spouts excited the imagination of men, making them curious and arousing their lust for adventure. In contrast, the small oyster hid under the water motionless and had to be discovered. The white man living near the Sound quickly discovered that the area abounded in that palatable food of the sea, and he went out in search of it, often garnering rich harvests. In the colonial period oystering was only part of the general search for food conducted by people along the shores of the Sound in order to supplement their diet. As a full-time occupation, regular oystering appeared only when markets developed ready and eager to absorb this delectable product.

Regular gathering of oysters began early in the nineteenth century in the Great South Bay region on the South Shore of Long Island. From here come the terms "bayman" and "oysterman" and "Blue Point" (a segment of the shore line near Patchogue where oysters of rare quality were found). The baymen operated with a simple tool similar to two rakes hinged together in the lower middle part of their long handles. Opening and closing the device, the operator could scoop up the oysters from the bottom of the shallow bay. The tool was clumsy, the work slow and tiresome, but many people were quite willing to make their living this way because oystering assured them a fairly good and steady income, and besides it gave them a rugged sense of independence. Piling up the catch of the day in his boat, the bayman would either sell it at the wharf to traders or open the bivalves himself and peddle the contents in the nearby villages. Sometimes people went to the wharves and bought the oysters fresh from the boats. As the industry increased because of the spreading popularity of the succulent product, packing houses were established at the centers of oystering, where the bivalves were opened or shucked and then packed in wooden kegs to be sent to New York and other markets. Stimulated by the great demand, the good oyster beds, discovered early, were soon exhausted, and so the search went on for new areas. The baymen did not have to go far, because Long Island Sound proved to be one of the richest "mines" for oystering. Great beds were found on both shores, especially in Oyster Bay, Huntington Harbor, Hempstead Bay, Smithtown, and on the mainland side in Norwalk, Bridgeport, and New Haven. But meanwhile new developments were taking place.

Unlike whaling, oystering could be a mass production enterprise. A large boxlike scoop, pulled on the bottom of the sea by chains or ropes, operated from a sailboat with the help of a windlass by one or two men, could do the work of ten to twenty hand-operated tongs. But the individual oysterman faced extinction unless he went into "farming" oysters, that is, planting oyster seeds and harvesting oysters four years later. For this he needed an area of several acres, a need leading inevitably to conflicts of interest. The public authorities were initially at a loss as to how to regulate this new trade. Regulation, as often happens, followed events instead of anticipating them, and events followed each other in quick succession. As the good beds were worked to exhaustion so that high-quality production could not be maintained, the quality of the catch declined markedly. An eyewitness, Timothy Dwight, observed in 1823 that Long Island oysters had "become lean, watery and sickly. . . . Formerly they were large and well flavored, now they are scarcely eatable and, what is worse, there is reason to fear, that they soon will become extinct."[15] Dwight's fears might have been justified had it not been for the switch-over to oyster farming. So complete was this transformation that by 1886 Eugene G. Blackford of the New York State Forest, Fish and Game Commission noted: "The oyster industry is rapidly passing from the hands of the fisherman to those of the planter and the oyster culturist."[16] Blackford concluded that the only hope for the oyster industry was cultivation.

Cultivation required careful preparation of the beds and their seeding with high-quality spawn which, at first, had to be imported to the Sound from Chesapeake Bay. Gradually cultivators around the Sound developed their own methods, so that by the end of the century seeds were sent from here to all parts of the world. Indeed, Long Island Sound proved to be an excellent field for oyster farming, largely because of the pioneering efforts of Connecticut.

Beginning in 1880 that state accorded privileges to towns on the Sound to allot areas within their boundaries to private persons for oyster farming. This power was exercised by oyster committees established by the towns. At first the allotments to each person were only two acres, which proved to be too small, and without police supervision it was impossible to protect the beds. This led to wholesale fraud and piracy; but new regulations brought order into the oyster business. Later Connecticut had the waters of the Sound surveyed by the United States Coast and Geodetic Survey, which fixed the boundaries of all holdings on a large map published by the State Oyster Commission. The amount of land was no longer limited, and the police patrolling the areas in boats protected the oyster farmers with buoys clearly marking the borders of individual holdings. With these regulations, the oyster business in Connecticut developed very vigorously, serving as a model to other states and especially to those parts of the Sound situated in New York State.

But other problems cropped up with all these new developments. In the course of two investigations in the New York area, Eugene Blackford noted the deterioration of the Saddle Rock beds near Great Neck. Of this area he said: "Years ago this bed produced large quantities of marketable oysters of excellent quality. The record of my recent investigation of this bed shows: Dredged seventy-five yards, found a roller skate, bottles, ashes, pasteboard, refuse. . . ."[17] At Execution Light, Blackford found a considerable amount of garbage in the Sound. Elsewhere the state had taken steps to eliminate this problem, and Blackford reported with justifiable pride that in Hempstead Harbor "an oysterman expressed much pleasure . . . over the arrest and conviction of parties violating the law against dumping garbage."[18]

Some enemies of the oyster could not be handled by legislation, however. Thus, Blackford observed that in Port Jefferson harbor the oysters were being attacked by their natural predator the starfish. Elsewhere in the Sound, at Throgs-Neck, Pelham Bay, Hart Island, and City Island, the oysters were thriving. In fact, the City Island bed was keeping oystermen fully employed, but farther up the Sound conditions were not so good. As industry spread, especially on the Connecticut shore, and contaminated the waters with its sewage, pollution became the bane of oystering. In Norwalk, for instance, one of the most important centers of oystering, the state had to put a ban on eating oysters from the cultivated beds of the town. In fact, the oysters had to be removed after two years to cleaner waters for the final growth to maturity. At the end of the last century, cleaner water could still be found on the North Shore of Long Island, where a thriving oyster trade grew up.

Private cultivators carefully studied the life of the oyster from spawning to maturity so that optimum conditions could be obtained or created for its growth. Thus, toward the end of the century, some of the artificial oyster beds could be extended into deeper water far from shore, some of them reaching out to a distance of six miles, with seeding beds located at depths of twenty to eighty feet. The technique of harvesting also developed at a pace with cultivating and farming. The first steam-driven dredge was introduced in 1874 in Norwalk, creating great derision among the oldtimers, just as the first railroad trains had been laughed at by the stagecoach operators, but irresistibly progress marches on. Toward the end of the century, giant steam dredges would reap rich harvests from fields deep under the water, unloading at the end of a day's work more than eight thousand bushels of large, meaty bivalves, the product of scientific farming. Large packing houses would process them, and distribution centers in New York and New Haven sent them to all parts of the country as well as to Europe. In fact, so efficient were the operations of these large oyster firms on the mainland side of the Sound, mainly in Connecticut, that they soon dominated the field, invading even the Great South Bay at the beginning of the twentieth century.

These large-scale operations did not, however, drive the small oyster farmer out of business. The old type of bayman, to be sure, could not compete with the dredge operators unless he went oystering as a private person to satisfy his own needs—provided he could find free beds for such sportsmanlike activity. Actually, ownership of small operations within the territorial waters of the towns remained high after the two-acre limit had been abolished, and even in the open waters of the Sound—that is, outside the town limits—there were many holdings of less than fifteen acres owned and operated by individuals.

These were the conditions around the turn of the century, showing great progress from the time when the bayman ventured out with his primitive tongs to make a living by arduous labor; and progress did not stop with the giant steam dredges. In the middle of the present century, new machines appeared, equipped with suction devices capable of gathering fifteen hundred bushels of oysters an hour. They replaced the dredges, which could do the work of three hundred men equipped with the ancient tongs. Now the new suction machines could do the work of four dredges—that is, of twelve hundred baymen; but what about the size of the oysters they gathered from scientifically cultivated beds seeded with eggs produced in laboratories? For the sake of comparison, the wild oysters of the tongs period, when shucked, measured 750 or more to the gallon. The cultivated oysters of the midcentury, when shucked, measured 170 or less to the gallon.

When discussing improvements in quality, we should note that the man responsible for investigating the oyster industry in New York State in the 1880s, E. G. Blackford, was also involved in improving the yields of fish in Long Island Sound through his work with the State Fish Hatchery at Cold Spring Harbor, Although the hatchery's saltwater activity ended in the 1920s, during its heyday the quality and quantity of fish in the Sound increased.

In the old whaling community of Cold Spring Harbor, the seat of the hatchery, clams were a local specialty. The area on the east side of the harbor bore the name "Clamtown." Although many clams were eaten locally, some were shipped by rail and steamboat to New York. It was a similar case with the lobsters taken in the Sound. Many were shipped to New York City, although some of the lobsters caught in Gardiner's Bay were sold in New Haven. Scallops were a specialty off Norwalk, while the creek which flows into the Sound at Mattituck on the North Shore of Long Island yielded delicious crabs. With the coming of steamboats and railroads in the nineteenth century these fruits of the Sound could be easily transported to the teeming markets of the metropolis beyond Hell Gate. Previously, only those who lived near the Sound could reap the harvest of the sea, but now those who dwelled in distant New York could enjoy its succulent delicacies.

Industrial Developments

Since the 1600s many changes had come to the Sound shore communities as well as to the waterway itself. These changes were especially apparent in the nineteenth century with the advent of the factory system on the mainland side of the Sound.

Industrial development had its beginnings just before the outbreak of the Revolution. Before that time it was in a primitive stage, supplying only local needs. The New England farmer, trying to squeeze a living out of the rocky soil, had to be versatile and resourceful. To keep his farm going and to provide the basic necessities for his family, he had to be a carpenter, a mechanic, or a blacksmith—even a mason if need be. At least, he had to produce enough surplus in order to be able to trade goods for services. Poor communications threw him on his own resources and those of his immediate neighbors. In short, to use a modern expression, he lived

76

in a do-it-yourself economy.

The long winter months were the time for tinkering in the barn or shop and for experimenting. If he had a forge or a lathe, he could fill orders from the outside and increase his income. Such people even developed special skills. Eli Whitney's father, for example, made chair posts on his lathe. His famous son, who set up a forge, hammered out nails and made hatpins for the ladies, who had abandoned the habit of tying down their bonnets. The skill which he developed in this way served him well later in making the wire teeth for his cotton gin. There were many such would-be inventors who, when the wars with England were over, could develop their talents and energies in a more prosperous world and launch the rapid industrial growth of the whole region.

An important factor in the development of industry in this area was the Sound itself. The waterway acted as a natural highway to and from the ever-growing market of New York City. From there the market boats sailed to points along the Sound where they could load up with the produce of the farms. The mouth of the Mianus River in Connecticut was such a point where the tidewater reaches up into the interior.[19]

On the arrival and departure of the market boats it [Mianus] was the scene of great activity with its crowd of farmers with their loads of produce, who purchased their supplies at one of the general stores. The market boats also made connections with the stages for the North and East.

Another center of activity developed closer to the Sound at Horseneck, later the borough of Greenwich. It attracted more and more business from Mianus and Cos Cob and became incorporated as a town in 1854. Here the Agricultural Iron Works turned out bar iron for tires, wagon wheels, horseshoe nails, rods for axles, etc. When the Civil War broke out, the company specialized in spike iron for the railroads and gun carriages. After the war, when competition from the large iron works became too severe, the works shifted to the manufacture of wire hoops for skirts, which were then the height of fashion. One branch of the factory also made spokes for carriage wheels and ax handles as long as the supply of hickory lasted. When it gave out, the mill closed.

As raw materials began to dwindle, some industries moved westward to be near newer sources of supplies. Economic dislocations in many New England towns resulted. Derby, for example, having access to two broad valleys, the Housatonic and the Naugatuck, and situated at the head of tidewater, enjoyed a favorable geographical position, attracting many factories. The Atwater-Hawkins steel works, built in 1847, employed about one hundred men. This factory grew beyond the ability of its waterworks to supply enough driving power, so the company decided to install steam. The fuel for the new power system was anthracite coal brought to the company's docks by schooner from Pennsylvania. But then came the rapid change when iron and steel manufacture shifted farther west where iron and coal were available in huge quantities for mass production, and the Derby mill had to close.

The shift of heavy industry westward was caused by the railroads. They brought about many dislocations, some of them very beneficial for towns and cities around the Sound. In some cases the railroads transformed the economy of the hinterlands, bringing great prosperity to the valleys debouching into the Sound. The Naugatuck valley was especially affected. It became a little Ruhr, alive with busy shops, plants, and factories.

Eli Whitney and the Birth of the "American System"

Along the Sound coast of Connecticut, there were many outstanding developments. In both Bridgeport and New Haven, deep-water harbors and the ease of transit for raw materials and finished products attracted many industries. In New Haven Eli Whitney changed the whole nature of the American factory system by introducing a production method utilizing machines with such precision that the component parts of the final product were identical and interchangeable. This system was born on the shores of the Sound when Eli Whitney in 1798 obtained an order from the government to produce ten thousand muskets in his shop in New Haven. Going to the national capital three years later, he had the opportunity to demonstrate his rifle to President Jefferson, who spoke with great enthusiasm of the high quality of this weapon. The president had acquired considerable knowledge while he was American ambassador in Paris. Himself a man of many talents and greatly interested in the mechanical arts, he had approached French inventors with a view to inducing them to come to this country and introduce the manufacture of firearms of the highest quality. Having failed in this attempt, Jefferson showed great enthusiasm for Whitney's accomplishment. He wrote later that the parts of the lock of the musket were so exactly alike that, if you took apart one hundred of them and intermingling them all, you could choose what you needed at random and put together a new lock without difficulty.

Whitney received additional orders, and the fame of his busy shop in New Haven spread far and wide. Visitors began to arrive, and the inventor himself entertained hundreds of them. Whether tourists or experts, they marveled at the machinery driven by water power which performed complex operations, turning out parts of the musket identical to the smallest degree. The advantages of interchangeability were obvious, except that they could be profitably applied only to mass production techniques. In Whitney's time most manufacturing processes were still in the stage of the shop run by the owner employing a few apprentices and workers. But Whitney was far ahead of his times. He laid the foundations by

78

Execution Rock Light Station (Official U. S. Coast Guard photo)

Old Field Lighthouse, now a private residence (Photo by Allan M. Eddy Jr.)

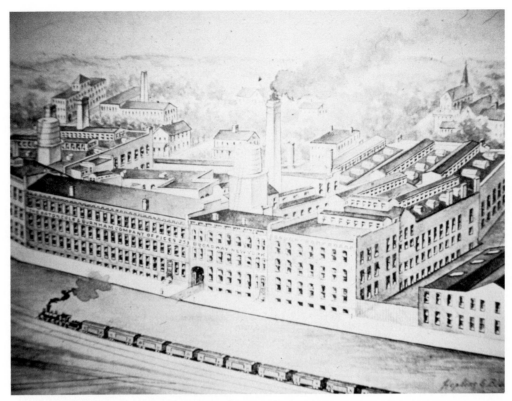

Eaton, Cole & Burnam, Co. (Fairfield Historical Society)

City Island Monorail (Hufeland Library)

setting forth a principle and demonstrating its usefulness. Slowly Whitney's ideas spread until they permeated the manufacturing methods and techniques of the American economy, becoming the great foundations on which, at its culmination a century later, Henry Ford could initiate a new industrial revolution by putting Whitney's method on the assembly line.

There were a few lines of production, however, which could adopt Whitney's system immediately. Sam Colt ordered from Whitney the special machinery he needed for the manufacture of his famous six-shooter. In a similar fashion clock works, sewing machines, and shoe factories, all in New England, followed the path indicated by Whitney's successful efforts. The manufacture of sewing machines could very easily adopt itself to this system. Although in the popular mind the development of the sewing machine is attributed to Isaac Singer, there were many inventors who, at the beginning of the last century, were experimenting with the idea of mechanizing an operation which hitherto had consisted of a needle, a thread, and the fingers of the housewife attaching two pieces of cloth to each other. It was slow and tedious work that could be sped up a hundredfold if done by a machine. This became a challenging idea to many inventors. In 1851 such a machine was exhibited in New York, attracting great attention. Nathan Wheeler, a light-metal manufacturer, decided to enter into partnership with the inventor and set up a factory in Bridgeport employing twelve hundred people, an unusually large number in the mid-nineteenth century. The reason was the breaking down of the operations into small details so that identical, interchangeable parts could be turned out by special machines. In one room there were "1003 separate machines for special mechanical operations, many of them seemingly possessed of such intelligence and skill as to direct and control their own movements, and only needing consciousness to rise to the plane of the skilled mechanic."[20] This was the American system of Eli Whitney developed as far as the technology of the period permitted it. There were, however, constant improvements. "No development of these works," continued the observer "is more interesting than the mechanical process of converting steel wire into perfectly finished needles. The distinct operations in the making of the needles now number 33 having been recently reduced from 53 by improved machinery."[21]

Years after Eli Whitney made his contribution to the American system of mass production, another man appeared in New Haven whose name became a household word in the American West and in sporting circles all over the world. He was Oliver F. Winchester, a shirt manufacturer of considerable wealth. His factory produced shirts by the so-called putting-out method. The material for the shirts was cut up in the factory according to various patterns and the parts distributed for sewing among households in the area around New Haven, although some of the shipments were sent across the Sound to villages on Long Island. The method was slow, however, and the product far from uniform, so that the shirts were often ill fitting.

The Winchester and Mass Production

A sudden change occurred in the shirt-making process when Wheeler paid a visit to Winchester. At first the visitor could not convince the host of the importance of mechanizing the process of shirt making until Wheeler's wife herself gave him a demonstration. Then he saw and believed. Soon five hundred pedal-driven sewing machines superseded the putting-out system, and Winchester's factory became a mass-producing organization. What he learned here was soon to be applied to the manufacture of firearms, and thus was born the famous "Winchester" whose shots were soon to be heard around the whole world. Although the rifles themselves were manufactured in New Haven by the mass production technique, the Winchester was not suitable for mass use. The soldier, nervous in the heat of battle, could easily cause it to jam, whereas if handled with great care it could be a superior instrument.

Clocks, in contrast, belonged to a different category. They required no handling beyond being wound by hand. Clocks, therefore, became very popular items, and the technique used in their production became world-famous. Indeed, by the middle of the century so great was the fame of the American method of mass production that in 1854 the British parliament sent a special committee to study the use of machinery in the United States. In New Haven the commission visited the clock factory of Chauncey Jerome, the Eli Whitney of clockmaking. In his plant 250 workers turned out 600 clocks a day. This was made possible by mass-producing interchangeable parts by mechanical means, so that three men attending special machines could produce all the wheels for the daily production of 600 clocks. So efficient was the factory that half of its products were exported to England, which at this time prided herself on being the workshop of the world. This competitive ability of American industry, as the commission stated, was due to mass production by the widest possible use of machinery. The committee found the same method used in the Davenport and Mallory works where two thousand padlocks were made daily. They could be sold so cheaply that a dozen of them could be had for fifty cents.

Carriages belonged to a different branch of manufacturing, where individual care and maintenance were important, but even here Yankee ingenuity could apply mass production techniques. With the opening of new turnpikes, there was an ever-widening market for carriages. The greatest number of buyers were in the South, where there was a market well organized for distribution by New York houses. Thus, the Southern trade and New Haven's proximity to its center in New York City spurred carriage manufacturing so that by the middle of the century there were over sixty establishments in New Haven turning out vehicles of all types.

One of these factories was able to complete a finished product hourly, extending its market beyond the South as far as Brazil.

In this American system low price was not the only result. High quality could also be produced, in some instances even the highest. While the padlocks turned out in the New Haven factory were for average use and available for the lowest budget, the Yale and Towne Company in Stamford produced high-quality locks, acquiring a reputation which went far beyond the boundaries of the United States. The nearness of one of the largest trading centers of the world, New York, plus cheap transportation by water, and finally the application of the American system could accomplish the task of mating mass production techniques with high-quality precision.

Yankee Notions and Hatting

The road toward this technological perfection was slow, and its tribulations, as we have seen, reached far into the past. Necessity was certainly one of the factors producing so many small shops and businesses in the valleys of the rivers emptying into the Sound. The number of shops and experiments increased. There were many small factories in Connecticut, and there was great industriousness in producing a large variety of articles and in an indefatigable searching for new ones. "Yankee notions" became a well-known expression to indicate the inventiveness of people who inhabited the bleak and rocky soil of this land. And there was a fluidity of industrial activity, though on a small scale initially. Little factories changed hands again and again, and here and there the fortunate ones would hit upon some new notion or some new method for which the time was just right, and a booming factory would go up. A few of them survived for decades or generations.

The hatting industry proved to be one of these flexible trades adjusting to changing conditions. The industry began on a small scale, with wide marketing done by peddlers. As marketing methods improved, a remarkable concentration of hat making took place in Danbury. From here it spread down to the Sound, centering in Norwalk where some very large establishments grew up employing hundreds of workers tending highly complicated machinery. By 1850 these factories could turn out a hat every minute, whereas at the beginning the quota had been three hats a day per worker. Increasing concentration and mechanization forced some of the manufacturers to lower the wages of their employees, so that the hat industry was often plagued by large-scale strikes, as in 1884–85 in Norwalk when there were riots and destruction of property. In 1909 an even larger strike spread in the industry. In Norwalk alone four thousand employees were affected. Strikebreakers brought in from the outside provoked street battles, causing large-scale migration of workers to Philadelphia and Danbury.

81

Thus Norwalk lost out in the bitter struggle, but this did not signal the permanent demise of industry there. That city and numerous others on the mainland side of the Sound, including the Westchester County town of Port Chester with its iron foundry and a host of other industries, Mamaronneck with its carriage factory, and the growing city of New Rochelle, were scenes of bustling economic activity in the nineteenth century. This was in sharp contrast with the situation on the other side of the Sound. There, with a few exceptions such as the shipbuilding industry in Port Jefferson, the pottery works at Huntington, and the woolen textile industry of Cold Spring Harbor, industry had made no significant inroads prior to the twentieth century. The North Shore of Long Island was still the preserve of the farmer and the fisherman, but before long all this would change, and Long Island, like the mainland side of the Sound, would be swept into the twentieth century.

Residence and shipyards of Jesse Carll, Northport Harbor, L. I. (Harper's Weekly)

6

The Dawn of Elegance
1900-1919

LONG ISLAND SOUND, WITH ITS LONG SHORELINE ON THE MAINLAND AS WELL AS ON Long Island, was bound, after its discovery by the white man, to invite settlements depending upon water transportation. The colonists established many villages and towns on the Sound before they proceeded inland. The roads radiated from seaports into the interior before lateral land communications between the shore settlements came into being. There was really no need for the latter. Sailboats could travel easily from harbor to harbor without burdening the inhabitants with the high cost of maintenance connected with highways. Like the people living on the shores of the Mediterranean in ancient times, the settlers of the "American Mediterranean" developed maritime habits. Depending so much upon the sea, their way of life and outlook were deeply influenced by it.

Even after steam revolutionized land transportation, making it fast and comfortable, travel by steamboat on the Sound remained fashionable. Only the arrival of the automobile brought about a radical change, a real transportation revolution leading to the disappearance of the majestic and luxurious passenger boats that plied the waters of the Sound for more than a century.

The First Horseless Carriage

In view of this lack of landward orientation of the people living around the Sound, it is a curious coincidence of history that one of the first inventors of the horseless carriage conducted his experiments on the shores of the Sound at Peacock Point, Long Island. The residents of the area must have looked with stunned curiosity and astonishment when they first saw Richard Dudgeon's steam-propelled vehicle being driven around on the bumpy roads of the North Shore, terrifying their horses. The inventor built his first vehicle in 1827, a rather strange looking contraption with steel tires and a smoke stack. It was roomy enough to seat ten, but there were disadvantages. On long trips it had to be accompanied by a wagon carrying cannel, a special kind of coal burning with a bright, hot flame. In spite of this complication the vehicle was quite an achievement. Its inventor, after winning a bet made with two other men that he could build a horseless carriage, had the joy of seeing his handiwork on display at the Crystal Palace. Fire in the Crystal Palace destroyed the original steam vehicle, but the new model Dudgeon built was still going strong in the 1920s. Across the Sound in Bridgeport the mayor and chief of police gleefully took a ride in it, imitating Fernando Wood, mayor of New York City, who many years before had tried it out.

Dudgeon's invention, though immensely popular in some circles, failed to inaugurate a new era in transportation. That would have to wait until after the turn of the century, when the gasoline-driven car came into widespread use. But even then the transportation revolution would be slow in starting. It would be the second decade of the twentieth century before the automobile was taken seriously as a method of getting from one place to the other. In the meantime people regarded the car as just another toy for the upper classes. Nowhere was this more evident than on the North Shore of Long Island, where the elite of New York City were carving out palatial estates, equipping them not only with the customary yachts, polo ponies, and racing thoroughbreds but with automobiles as well.

The Coming of the Automobile: The Vanderbilt Cup Races

William K. Vanderbilt, one of the earliest devotees of the automobile, became involved with cars after his interest in golf, which he helped organize in this country, apparently declined. His patriotic soul felt alarmed because at that time only the French seemed able to build good automobiles. In an attempt to encourage American auto manufacturers to improve their product, Vanderbilt introduced an annual sporting event on Long Island known as the Vanderbilt Cup Race. These quickly became popular from the time of the very first contest, held on October 8 1904. Although an Englishman behind the wheel of the French car going at the staggering speed of fifty-two miles per hour won the cup designed by Tiffany and Company, within two years an American driving an American car captured first prize.

At the beginning the cars ran on public roads in Nassau County. At dangerous curves barriers kept the spectators out of the road until the racers were approaching. Then the enthusiastic crowds, eager to get a close view, crashed through the barricades with disastrous results. To prevent further accidents, Vanderbilt decided to build a private road, constructed without sharp bends and with ample width so that the spectators could be better controlled. Difficulties with land acquisition and other problems delayed construction. But eventually, after the Vanderbilt Motor Road, as it was popularly called, became reality, motor racing grew into a big business requiring specially built circular tracks with grandstands. The Vanderbilt Motor Road, incidentally, caught the fancy of motorists, who loved to drive on it even though its owner charged stiff tolls; and lest enthusiastic drivers start unofficial races on it, speed limits were strictly enforced.

Before the completion of the Vanderbilt Motor Road, public roads continued to serve as the track for the Vanderbilt Cup Races. Despite the obvious danger, or perhaps because of it, the races were very popular. When the races were held, the Rye-Oyster Bay ferry had to put in special runs to take care of the swollen traffic. In 1910 the Long Island Rail Road arranged boat service across the Sound from New London, Connecticut, to Greenport, where racing fans boarded special trains taking them directly to the race. Four spectators lost their lives that year, causing an immediate uproar. Some of the drivers declared that they would not participate in future Vanderbilt Cup Races unless safety standards were improved. Insistent demands that the races be stopped forced the New York State legislature to pass a law banning races on public roads. A loophole in the law, however, permitted public roads to be used for races with the approval of local government. The Nassau County Board of Supervisors quickly voted permission, but Vanderbilt gave up the idea of continuing the races on the public highways. He planned instead to build a special auto racetrack at the end of his motor parkway at Lake Ronkonkoma. The coming of World War I made him put aside the project.

The Automobile and Suburbia

After the war, as the country entered the prosperous period of the 1920s, the automobile ceased to be merely the toy of the rich. It became instead one of the mainstays of transportation for the average citizen, and much of its glamor disappeared. As use of the auto became more widespread, a huge clamor arose everywhere for the improvement of existing roads and the construction of new ones. Even in the heretofore quiet little communities along the Sound, the din became louder as the twentieth century advanced. In 1913, when hearings were held on letting contracts to improve the highways in Nassau and Suffolk counties, residents demanded that good construction materials be used. Some of the witnesses were especially adamant about using something more substantial than Hudson River gravel as a base for the roads.

Such problems had not troubled the first generation of motorists back at the turn of the century. They did not use their imported vehicles so much for transportation as for pleasure drives in the vicinity of their estates. But by the teens of the twentieth century all this was changing. Prior to World War I there was already an army of exurbanites moving from New York City out to the North Shore of Long Island and northward into Westchester County. The majority of these people were by no means lower class, but neither were they the wealthy elite who could afford to build mansions along the Sound. They were middle class and upper middle class people, the first wave of suburbanites drawn to the North Shore of Long Island and the Sound areas of Westchester and Connecticut by the same attractions which had brought and were still bringing the rich. They were the first wave of what was to grow years later into the Big Sprawl.

The Gold Coast of Long Island

Before the arrival of this new class of people, however, the Sound-side areas were monopolized by older wealth whose origin went back to the nineteenth century when on both shores farms had given way to estates and waterfront commercial areas had been upgraded into residential property. Where crops once grew, manicured lawns and formal gardens blossomed. In the first years of the twentieth century, when driving cars was considered a sort of elegant sport requiring goggles and a linen duster, the rich monopolized the waterfront. Arriving at their Sound-side estates by yacht or train, they settled down to enjoy the good life for the weekend or sometimes for the entire summer. Along the North Shore in Nassau County the wealthy émigrés, mostly from New York, built magnificent mansions, transforming the area into a veritable Gold Coast.

This term, which is so indicative of the lifestyle of the upper class residents of the North Shore, is also used to describe an era in Long Island history, the period approximately between 1900 and 1920, when the great estates flourished. Handsome houses had been built on the Sound prior to the turn of the century, but there was something very special about the palaces erected in the early 1900s. They were usually much larger and more ornate than their predecessors; they became even more beautiful or gaudy, depending upon one's viewpoint, as their owners vied with one another to see who could create the most splendid estate. Typical of the Long Island mansions of this era was the one built in 1903 by Louis Comfort Tiffany overlooking Cold Spring Harbor. Called Laurelton Hall, the house had eighty-two rooms and twenty-five baths. Vice president and director of Tiffany and Company, New York, Louis Comfort Tiffany was also an accomplished artist. Indeed, at the present time glass designs executed by Tiffany are eagerly sought by antique collectors and when available bring exorbitant prices. Exorbitant might also be the word

86

to describe the $13 million spent by Louis Comfort Tiffany's son Charles, a diamond merchant, to refurbish his father's estate.

Even without remodeling expenses the cost of keeping up a huge house was high, though not really prohibitive in an age when servants were cheap and income tax nonexistent or diminutive. J. P. Morgan once said something to the effect that anyone who had to think about the cost of maintaining a yacht should not get one. Morgan surely knew what he was talking about, because the upkeep alone for the different *Corsairs* which he had must have been quite high, considering that even the first *Corsair* had a crew of sixty. Given the depth of the Sound, maneuvering such a craft was no problem, and since many of the estates had oversized docks capable of handling even the largest vessels, the yachtsman could drop in from the waterside to see some of his fellow estate owners.

For many of the Long Island elite, property on the Sound was not the only prerequisite for a magnificent estate. In addition to having water rights, the land also had to have a view. Thus, many of the Gold Coast mansions, as well as those erected on the Westchester and Connecticut waterfronts, were built on cliffs overlooking the Sound. Sometimes the estate owners erected other structures on high land, too, such as the Temple of Love which H. L. Pratt of Standard Oil installed on his estate at Glen Cove, offering a breathtaking view of the Sound.

Temples of Love

Some of the big houses on the estates were themselves temples of love, as for instance the home built by Jay Gould's son Howard at Sands Point and later sold to Solomon Guggenheim who lived there during the height of the Gold Coast era. In 1898 Howard Gould married Viola Catherine Clemmons, an actress, despite the objections of the Gould family, none of whom attended the wedding. Since the new Mrs. Gould liked Kilkenny Castle, her husband engaged an architect to design a replica of that structure. Howard Gould even sent the architect to Ireland. But when he returned after several months, he was unable to duplicate the castle at Sands Point. Viola was unhappy, so her Howard had the stable, known as Castle Gould, modeled after Kilkenny Castle. He also proceeded to sue the architect, but that was just the beginning of his legal entanglements. He and Viola eventually ended up in a court fight, with Howard accusing Viola of infidelity with none other than "Buffalo Bill" Cody. Shocking as this revelation was, it did not permit Howard to regain his freedom. Viola persistently refused a divorce and remained Mrs. Howard Gould until her death in 1930.

The marital difficulties of the Goulds, or for that matter those of another prominent Gold Coast resident, Clarence Mackay of Roslyn, whose first wife ran off with a doctor, made fascinating reading. People love to follow the careers, escapades, and scandals of the upper classes, deriving a vicarious pleasure from them. To satisfy this craving, the society pages were an important feature in every newspaper. But it was not only personalities and their deeds and misdeeds which attracted attention.

Mystique of the Gold Coast

A considerable interest generated by the North Shore in the early twentieth century emanated from the great estates themselves. The mansions' sometimes more than their owners had a certain mystique, and the newspapers delighted in describing the beauty of these estates without even mentioning the name of the owner. Until the beginning of the Great Depression, the real estate section of the Sunday *Times* abounded in news of large land purchases, and often when a famous estate changed hands the new owner was described simply as a man of means or a prominent industrialist, but his name was not given. Needless to say, there were exceptions to this rule. During World War I, for example, when the railroad tycoon Otto Kahn was building a mansion at Cold Spring Harbor, the *Times* clearly identified the owner and described his new home, really a French chateau, with 126 rooms, which was said to have cost $9 million. At one time the grand ballroom held two hundred people for dinner and the guests were tended by two hundred servants. The sum of $9 million was also the reputed cost of the Glen Cove mansion built in 1914 by one of the Woolworths. Farther east on the island at Oyster Bay, homes described as "simple Italian villa types"[1] were completed for other wealthy New Yorkers. In an era when Smithtown Bay was described as looking "not unlike the Bay of Naples,"[2] waterfront property all along the Sound was being snapped up. Whether it was the aura of personal wealth, prominence, or success — nothing succeeds like it! — or the mystique of fabulous palaces, the general lifestyle of these people shed luster on this area of the North Shore, making the very name "Gold Coast" synonymous with glamor and brilliance, like the word "millionaire" evoking instant associations of sleek yachts, sleek race horses, and sleek ladies.

A Paradise for Horse Lovers

Many of the new property owners along the Gold Coast invested in Sound shore real estate either because they liked the view or were enthusiastic about water sports, including yachting; but some North Shore residents had other interests. As far back as the colonial era, Long Island had been horse country, and races were run before the American Revolution. By the early twentieth century thoroughbred horses abounded in the Wheatley Hills area of Westbury, the immediate hinterland of the Gold Coast.

Since horse breeding was introduced in England in the eighteenth century, owning stables, horse racing, and riding horses for sport was the distingushing mark of the real gentleman. The Gold Coast with its hinterland became the true home of this sport and of the lifestyle which it inspires, a remnant of the feudal days when horsemanship was the prerogative of the upper classes as owning and driving an imported flashy automobile was later the privilege which money conferred upon the rich. Concentration on these elegant pursuits, categorized as conspicuous consumption in the language of sociology, produced on the estates of the Gold Coast the finest breed of racers, polo ponies, and fox hunters. Many racetracks and polo fields could be spotted here by the traveler, and even the country clubs specialized, some being the favorite haunts of the fox hunters, the polo players, or the racing enthusiasts. The area boasted of many horse shows held in Sands Point, Stony Brook, Brookville, Piping Rock, and Locust Valley, places which still evoke nostalgia among horse lovers.

The Gold Coast: A Lifestyle

In this region there were mansions which seemed like baronial estates moved straight out of eighteenth-century England. Stable boys busied themselves around their charges, sheep grazed on the nearby meadows, the bell hanging around the ram's neck chimed a gentle tune, a pond waited with water lilies and boats ready for a silent glide over the placid waters—all a perfect setting silently yet eloquently proclaiming that this was the home of the elite who filled their abundant time with truly aristocratic pursuits. Thus, the Gold Coast acquired a special place in the popular mind, its name a special ring. It had no clearly defined boundaries, and one would have searched for it in vain on the maps. As long as the estate overlooked or was closely located to the Sound and its lifestyle observed the highly sophisticated standards of the elite, it might be termed part of the Gold Coast. With this loose definition, the boundaries could be extended to the other side of the Sound. In Larchmont, for instance, Rudolph J. Schaefer, who had accumulated a fortune in a business soon to be legislated out of existence by the Prohibition amendment, purchased the waterfront mansion now serving as the clubhouse of the Larchmont Shore Club. The house was described as being of "perfect Elizabethan architecture,"[3] with practically every room facing the water. One of the finest estates on the mainland side of the Sound, La Hacienda, as it was called by its former owner Mrs. Aimée Crocker Gouraud, had a fifteen hundred-foot waterfront, several gardens, stables, tennis and squash courts, and greenhouses.

Spread of Suburbs

In the first two decades of the twentieth century, communities on the mainland side of the Sound, as on Long Island, were attracting not only people of great wealth but the moderately wealthy too. Numerous tracts of waterfront land as well as inland parcels in the Sound shore communities were bought by New Yorkers for summer homes and increasingly for year-round dwellings. Westchester County, particularly, attracted this first generation of suburbanites because of its proximity to the city. In 1913 the *New York Times* observed that "grand historic Westchester is waking up."[4] The county once had the reputation of being unhealthy, but on the eve of World War I potential residents were not deterred by the old stories of malaria and typhoid fever. Indeed, in 1913 the New York State Department of Health rated one Westchester Sound shore community, New Rochelle, the healthiest city in the state. In an attempt to generate additional favorable publicity for Westchester, its promoters organized a "Country Life Permanent Exposition" for Grand Central Station to convince New Yorkers that Westchester County was the place to live. The exhibit displayed a model home equipped with the latest appliances, and five scenic paintings, including one of Long Island Sound at Oakland Beach, Rye.

As a promotional event, the exposition naturally failed to reveal that everything was not pleasantly bucolic in Westchester. Advances of every type, from a new public library and post office in New Rochelle to an improved Boston Post Road, were taking place; at the same time an increasing tendency prevailed to exclude everyone but the middle class—this is, its upper levels and the extremely rich—from scenic Westchester. A newspaper article dealing with the cost of living in the county in 1915 pointed out that although some people actually spent less money by moving to Westchester, such as the man who saved five hundred dollars on doctor bills after he bought a house in the "country," one had to have a fairly good income to afford a home in the $10,000–15,000 bracket. The buyer of a ten-room house with three bathrooms might be getting a lot for $15,000, at least by our standards, but he had to pay off the mortgage at a time when $4,000 constituted a good income and most workers earned considerably less.

For those with small incomes homes in the $4,500–5,000 price range were available in a new development at Flushing Bay, Long Island. In the words of its developer, this community was ideally suited for "clerks, bookkeepers, and other city workers of a cultured type who are anxious to live in the country with pleasant surroundings and in an agreeable neighborhood, but whose means to enjoy these facilities are limited."[5] As one of its big attractions, in addition to tennis courts and bath houses, it granted each home owner perpetual rights to use the beach.

Speculators, Developers, and Dreamers

Elsewhere on Long Island people of moderate means who invested in property were not so fortunate. During the first decade of the twentieth century, speculators bought up large tracts of barren land in Suffolk County and resold the property, sight unseen, to people throughout the United States. Salesmen of the land companies made commissions of as much as 30 percent, and some are said to have earned two to three hundred dollars per day. The purchasers of the land often did not learn that they had made a bad investment until they attempted either to move to their property or to sell it. In the latter category were a number of investors from Dayton, Ohio, who, when a flood hit their city, tried to sell their Long Island property and learned only then that they had purchased worthless land. The unscrupulous activities of the land sharks prompted one reader of the *New York Times* to write to the editor, "My advice to all is to leave L.I. property alone."[6]

But this was not altogether good advice, because in many parts of the island land was steadily increasing in value during the first two decades of this century. By 1917 many of the farms along the North Shore had been broken up and sold at prices which would have been considered absurd at the turn of the century. The same thing was happening on the mainland side of the Sound, but at a slower pace. On the eve of World War I, developers had made few inroads. The shore line from Larchmont to Stamford consisted of big estates, but here and there developers had driven an entering wedge. Shippan Point, jutting out into the Sound at Stamford, was being built up, and at Southfield Point a dozen houses in the $10,000–50,000 class were constructed.

Across the Sound on the Great Neck peninsula, developers were making inroads. They created a number of new communities, among them Kensington, an exclusive suburban development with entrance gates being an exact replica of those at Kensington Park, London. The elegant community had many innovations to offer residents: concealed utility lines, ample park land, and waterfront facilities. Moreover, the homes were 150 feet above sea level, thereby "practically eliminating mosquitoes, malaria and other objections to water level locations."[7] A promotional pamphlet describing life in Kensington tempted potential residents with "sunny days filled with outdoor activities followed by a quiet evening on the veranda, where you can enjoy to the full the cool Sound breezes and the quiet of the country nights."[8] In the Manhasset area the construction of the Plandome Estates followed shortly after Kensington was created. As a big innovation, completely furnished model houses tempted the prospective buyer. In its immediate vicinity rose the Hempstead Country Club, boasting one of the finest golf courses on the North Shore. It had two thousand members and a waiting list by 1916. Farther east on the island the Glen Head Country Club offered the dual attractions of golf and boating. An early club brochure described the excellent features of the land which made it adaptable for a golf course, and the advantageous location of the club's boat landing in the heart of the Sound's yacht racing area just across from Larchmont.

Members of fashionable clubs on both sides of the Sound sometimes used their yachts for commuting back and forth to New York City as well as for weekend jaunts around the Sound. But already the teens of the twentieth century made it increasingly apparent that despite the proliferation of yachts, boat transportation on the Sound could not compete in popularity with the railroad for commuting to the city. Commuting by boat was largely a phenomenon of the nineteenth century, while auto commutation would not come into its own until the 1920s. In the interim period the railroad was king on both sides of the Sound. This did not mean, however, that new transportation methods were overlooked. Probably the most fantastic scheme conceived for the Sound area was the proposal to build a fourteen-mile long ship canal from Port Chester to Tarrytown in Westchester County. The canal, estimated to cost $3 million in 1906 when it was proposed by a group of Mount Vernon, New York, businessmen, would have linked Long Island Sound with the Hudson River. Judge Adam E. Schatz, who spearheaded this project, was the inventor of the "electric mule," a device for towing canal boats. According to him,[9]

with the canal from the Hudson to the Sound, Westchester would be surrounded by water on all sides. It would become the Venice of America . . . the lower freight rates . . . resulting from the operation of the canal would bring on an era of tremendous growth in all directions; farm lands would be cut up for residence purposes, houses would go up everywhere.

Judge Schatz's scheme never saw the light of day, but another fantastic dream, a monorail from the Bartow station of the New Haven Railroad to City Island bridge, did. Constructed in the early nineteen hundreds, the monorail was the only one of its kind in the United States. But in 1910 a serious accident occurred, discouraging further development.

Trolley Lines and the Railroads: The Case of the New Haven

Canals and monorails notwithstanding, prior to World War I the railroad remained the preferred means of transportation on both sides of the Sound. The Long Island Rail Road provided improved service on the newly electrified North Shore division. In some areas street railways and trolley lines facilitated travel from one community to another. Street railways also abounded on the mainland side of the Sound where Charles Mellen, president of the New Haven Railroad, decided to acquire every competing line on land and sea. He made some serious mistakes, however, his worst being the 1907 decision to purchase the controlling interest in a trolley line not yet in operation but calling itself the New York, Westchester and Boston Railroad. Construction began in 1909 for the stretch from New York to Port Chester, and by the time operations started in 1913 the New Haven had spent $22.3 million on 18.3 miles of new track, or $1.2 million per mile. Investigation brought out indications of an additional $14 million expenditure on different and often mysterious accounts, so that the total cost came to $36 million all of this spent on a line actually competing

92

with the New Haven since its tracks ran parallel to it. The investigations were not pursued any further, and the real reason for these obscure shenanigans was never uncovered. Suffice it to say that J. P. Morgan, the celebrated financier, the real power behind the management of the New Haven, was afraid of competition, for this electric trolley line, as its name indicated, might be extended to Connecticut, reaching perhaps as far as Boston. This of course would be a serious matter for the New Haven. Under Morgan's instructions the New Haven opened a special account in 1905 from which the promoter of the trolley line, an obscure figure named Oakley Thorne, could draw. He took advantage of this privilege, borrowing a total of $11.5 million in cash against a miscellaneous bundle of securities. When the panic of 1907 hit the money market, bankrupting Oakley Thorne, nobody could find out where the cash went. The securities left in his account were worthless, and Morgan prevented further inquiries into the matter. Subpoenaed by the Interstate Commerce Commission during the famous investigation of the New Haven Railroad in 1914, Thorne informed the committee that he had burned all his records before the Senate ordered the investigation and he could remember nothing.

In spite of the fraudulent and scandalous conduct of the promoter, the New Haven proceeded to build the trolley line. Plans were made to extend a branch from Greenwich to Danbury. Other branches of the line, one to New Rochelle and the other to White Plains, were completed in 1912 and began operations as rapid-transit railroads. In New York they connected with the Third Avenue elevated in the Bronx. From the beginning the New Haven lost money, some $4.3 million in three years. The bonds and interests of the trolley line being guaranteed by the New Haven, the coverage of the losses produced a continuous drain.

Investigation of the New Haven

The investigation of the New Haven Railroad at which Oakley Thorne testified had come about because on February 7 1914 the U.S. Senate had passed a resolution ordering the Interstate Commerce Commission to look into the management of the railroad. In almost continuous sessions lasting sixty days, it uncovered conclusive evidence of maladministration. Published by the Senate in two bulky volumes, the minutes of the hearings contain several hundred pages. All the records of the New Haven were under scrutiny, that is, those records that could be subpoenaed; even then they had to be brought into the hearing room in large clothesbaskets. Many others, including the books and much of the correspondence, had been burned or otherwise destroyed before the investigation started. To make matters worse, many of the witnesses refused to testify during the preliminary hearings until threatened with criminal proceedings. It was one of the most difficult cases to pursue, the investigators being compelled to unravel some of the most intricate legal devices used by the administration of the railroad to hide unlawful practices.

Charles S. Mellen, the president of the New Haven, had the infamous distinction of being the chief culprit of this sordid scenario. For five days he was on the stand being mercilessly interrogated and cross-examined. With great stamina and presence of mind, often demonstrating his sharp wit, he protested his innocence, putting the ultimate blame at least by innuendo on J. P. Morgan, who controlled the New Haven.

Only one man could clip the wings of this all-powerful director of American finance, and that was Theodore Roosevelt, the president of the United States, who forced Morgan to abandon his plans to establish a monopoly of the railroads in the whole northeastern region of the country. In view of the Supreme Court ruling against transportation monopolies in the Northern Securities case, Morgan had to proceed very carefully with his plan to create a new railroad monopoly for New England. His chosen instrument for this design was Charles S. Mellen. In attempting to circumvent government regulations, Mellen made some costly blunders, and ultimately the New Haven suffered its greatest losses.

Decline of the New Haven

At the turn of the century, the New Haven's stock was among the bluest of blue chips, and lucky owners considered themselves as belonging to an elite group. To be elected a member of its board of trustees was the mark of the highest distinction in the business world. No one questioned the soundness of its management or its continued profitable operations. Yet in little more than a decade after the turn of the century, all this prestige and reputation vanished, giving way to bitter disappointments and accusations followed by investigations. At the peak of its financial chart, New Haven stock was quoted at $216. From this height it declined to $49, causing widespread misery and ruin among the stockholders. American railroad history of the nineteenth century is full of misdeeds, speculation, and scandals, and those railroad men who committed these crimes could fill an impressive rogues gallery. Still, the story of the New Haven is unique in the respect that some of the greatest luminaries of the financial world were involved in a monstrous game to create a complete transportation monopoly over a whole region of the country without the interference of the public authorities on either the state or federal level. Charles Mellen, the man selected by J. P. Morgan to accomplish this feat, was amply supplied with money, and he poured it out rather freely. The report of the Interstate Commerce Commission stated that out of the total increase of capital under Mellen's administration $204 million was expended for projects outside the sphere of railroad operations and only $120 million was devoted to the railroad itself.

Indeed, a great deal needed to be done to modernize the New Haven. Many old wooden bridges, no longer able to carry the heavy loads of modern freight trains, had to be rebuilt. The rolling stock as well as many of the locomotives were obsolete.

94

But while Mellen busied himself renewing the old equipment, he was, at the same time, looking around to find actual or potential competitors whose property he could acquire. He used various means, and the report of the investigators tersely states that the New Haven suffered its greatest losses in attempting to circumvent government regulations and to extend its dominion beyond the limits fixed by law.

The New Haven and Shipping on the Sound

One of the principal objects of attention of the Morgan-Mellen campaign to build up the New Haven's transportation monopoly was the trolleys, but in addition to various forms of land transportation the directors of the New Haven Railroad had their eyes on the Sound steamers as well.

Before Mellen became president, the company had been operating several steamship lines. They were acquired in the nineteenth century when the railroad absorbed other roads which ran from the Sound into the interior and operated steamship lines of their own to complete the connection with New York. With all these lines, the New Haven could offer steamship connections from New York to Newport, Fall River, Providence, Norwich, Stonington, New Haven, and Bridgeport. The vast fleet of passenger lines, some of them equipped with the latest comforts and luxuries of the times, failed to satisfy the acquisitive instincts of Morgan, who was said to have taken great pride in posing as a lover of the sea. Mellen, who professed little interest in steamship lines and was much concerned about the hazards of storms, collisions, and fires, had to go along. Purchasing trolley lines made sense insofar as the ownership of electric rail lines guaranteed the absence of competition. No one in his right mind would build another trolley line alongside the old one owned by the New Haven, but it was different with steamship lines.

Anyone adventurous enough to take a risk might come along, buy a few passenger boats, and start a steamship line. Then he would wait to be bought out for a good price by the New Haven. Some owners of competing lines sold out at high prices, others continued the competition stubbornly. The New Haven, however, went on fighting back, building faster and more palatial ships to hold its own. The Sound had not seen such feverish building activity since the early days of Vanderbilt and Fisk. Nor was there ever more legerdemain in manipulating charters. There were many purchases and mergers so that the New Haven could hold its own over the waters of the Sound. If a competitor arose, he was sooner or later forced to sell or go into bankruptcy; but there are exceptions to every rule.

Morgan and Mellen Find Their Match

Charles W. Morse was a man of different mettle who for years, by hook or crook, defied the New Haven. A promoter with real genius, the "ice king" of New York, a name he earned by cornering the ice business of the city, went into shipping and founded the Consolidated Steamship Company, which controlled several other steamship lines. In 1905 Morse announced the building of two ultramodern passenger boats, the *Harvard* and *Yale,* powerful and large enough to sail around Cape Cod, establish a direct connection between New York and Boston, and avoid the necessity of transferring to rail at Providence or any other point. Operating the famous Fall River line since it acquired the Old Colony Road in 1893, the New Haven answered the challenge by ordering a new side wheeler, the *Commonwealth,* which, with the luxurious *Priscilla,* introduced in 1894, was to hold its own. A rate war and speed race ensued, with several consequences. One was that Mellen had to go to Washington to see President Roosevelt and assure him that the New Haven harbored no plans to secure a monopoly of shipping on the Sound. On the contrary, he pleaded with the chief executive, saying that if the New Haven liquidated its shipping interests, Morse would be in the position to establish a monopoly. The President, who disliked Morse and his sharp practices, agreed, but later became suspicious, seeing that the New Haven continued purchasing new lines. In February 1907 the *Larchmont,* a ship of the Joy Line, which had recently sold out to the New Haven, collided with a freighter causing the loss of 131 lives. More fortunate were passengers aboard the *Commonwealth,* which in 1908 while speeding in a thick fog ran into the Norwegian freighter *Voland.* No lives were lost in this accident.

Other problems beset the ventures. As a result of the panic of 1907, Morse's shipping lines went into receivership, and Morse himself was sentenced to a long prison term. President Taft later pardoned him, yielding to the importunings of Mrs. Morse, who asked for clemency in view of her husband's very serious illness; but the illness proved to be a fraud because Morse, expecting the visit of army doctors sent by Taft, drank soapy water of such quantities that he had temporary bleeding of the kidneys. But this convinced the army doctors. Free again, Morse resumed his nefarious activities, but the New Haven succeeded in acquiring all his remaining shipping interests, so that the Morgan-Mellen team now owned or controlled 90 percent of the water transportation on the Sound. During World War I the government took over all railroads and shipping lines, but in 1918 the Interstate Commerce Commission handed down a decision whereby the New Haven could retain its shipping lines, since it was assumed that shipping in the Sound could not continue without the New Haven's support.

Tragic Accidents on the Sound

In general, the steamboat business was booming in the early twentieth century for cargo and passengers alike but occasionally there were accidents. One of the most tragic involved the big Sound steamer, *Slocum.* Bound for the picnic place of Locust Grove with well over one thousand people on board, many of them women and children on a church excursion, the boat caught fire possibly because of an explosion in the galley. Although the *Slocum* had recently been refurbished, the life preservers on board had not been replaced, and when the old life jackets were donned, the straps, rotted by age, failed to hold, with the result that some of the women refused to jump into the deep water off North Brother's Island where the captain tried to beach the boat. Instead, they burned to death with their babies in their arms. The pastor of the church who sponsored the excursion went into shock so badly that he was not expected to live. His wife and two daughters were aboard the boat when the accident occurred. The bodies of his wife and one daughter were recovered, but the other daughter was missing. Investigation of the accident revealed that the *Slocum* had numerous collisions in the past and had been fined for overcrowding. Perhaps the saddest aspect of the tragedy was that a yacht followed the burning boat through the East River but sailed away, refusing to pick up any of the survivors who had been hurled into the water.

A survivor of the loss of the famous Sound steamer *Larchmont* in February 1907 off Watch Hill, Rhode Island, noted a similar instance of inhumanity. An explosion had forced crew and passengers to abandon ship. As one of the lifeboats began moving away, the people aboard noticed a woman floating by in the icy water. The man who recounted the story shouted to his companions to pull her in, but they refused lest she weigh down the boat.

World War I

With the outbreak of World War I, although the Sound was far from the danger zone, the fear of hostile action of some sort loomed in the minds of some area residents as the country geared up for war. As part of the military preparations, the government established training camps on Fishers Island in the Sound and at Hempstead and Yaphank on Long Island. At the same time many Sound shore communities became beehives of activity. Industrial centers like Bridgeport and New Haven turned out weapons and ammunition for the American Expeditionary Forces and their European allies. It was assumed that Connecticut would be especially attractive to attack from the Sound, owing to the presence of munitions factories in her major cities. The Connecticut General Assembly pondered this possibility even before the United States became involved in the war. But fortunately the legislators' dismal visions of a repeat performance of the Revolutionary War and the War 1812 failed to materialize. The assembly, nevertheless, authorized an inventory of the manpower and resources of the state. This precautionary measure, implemented prior to the declaration of war, was typical of the steps taken by Americans to prepare for the conflict thought by some to be inevitable.

F. Trubee Davison, a young resident of the North Shore, made one of the most unusual contributions to American preparedness. Impressed by the Lafayette Escadrille during a 1915 visit to France with his father, a partner in J. P. Morgan and Company, Davison formed the flying unit which proved to be the beginning of naval aviation. Gathering a handful of his Yale classmates, he began to experiment with a flying boat belonging to a Port Washington aviation corporation. For the young men in this first Yale unit, flying was a great adventure, and even near misses were exciting provided you lived to tell the tale. Once, while flying a training mission, Davison himself narrowly missed the Queensborough Bridge, and another time was lost in the fog on a flight from New Haven to his home at Peacock Point. On the latter trip he got his bearings when the aroma of Bridgeport industrial plants wafted into the cockpit. Davison regularly flew across the Sound and, on occasion, startled the Yale administration by going home to Long Island for the weekend and coming back to New Haven by plane in time for Sunday chapel. Finally his luck as a pilot ran out. While making a demonstration flight for the Navy in 1917, Davison's plane plunged to the bottom of Huntington Bay. Because of his severe injuries the first Yale unit had to leave without him for France. But in recognition of his having formed the first Yale Unit, the U.S. Navy in 1967 presented wings to the man regarded as the father of naval aviation.

Testing of planes over the Sound went on parallel with testing of submarines under the Sound. During World War I New London served as a submarine base, and it was from there that America's answer to the Kaiser's U-boats went out on test runs. When the United States actually became involved in the war, the Sound came to be used more and more for military purposes, and the activities associated with the waterway in peacetime were at least partially curtailed. The big Sound steamers were shunted from their customary berths in the Hudson to the East River. The new location was actually closer to Long Island and therefore more accessible to the many Long Islanders who used the boats, but the rationale behind the move was not convenience of the passengers but a desire to free the Hudson River docks for military and other vessels needed for the war effort. Another consideration may have been the fact that the Sound steamers berthing on the Hudson had to pass the Brooklyn Navy Yard on their way out of the harbor, thereby affording any spies on board an excellent opportunity to photograph the work going on in the yard. That this may have been the factor in the decision to move the boats is not so absurd as it may appear. Americans were very conscious of security, and nowhere was this more the case than in the New York area, where as early as 1917 the National Guard was patrolling all bridges and stopping cars which were deemed suspicious. The East River bridges were given high priority; five Coast Guard vessels were on patrol, and any private vessel tied up near the bridges could be seized.

Despite all the precautions necessitated by the wartime situation, in the more outlying sections along the Sound life went on as usual. For several years summer rentals declined, as people engaged in war work had to stay in New York or in Washington, and for a time housing starts were slowed by a shortage of building materials. However, in the more rural areas of Connecticut and Long Island things seemed quite normal. The only difference may have been that the war toned down the social life of wealthy Sound shore residents, as millionaires as well as many ordinary folk turned to patriotic activities, including gardening. Shortly after the United States entered the war, that intrepid Sound shore resident Theodore Roosevelt told the Long Island Farmers Club that the planting of flowers and the transformation of grain into alcohol should be curtailed for the duration of the war. Instead, everyone should plant food crops, everyone but Teddy apparently, because when the former president had finished his speech, Ralph Peters, president of the Long Island Rail Road, asked him: "Colonel, are you going to France or will you remain here and help us raise crops on Long Island?" T. R. responded: "By George, I'll go to France if they'll let me!"[10]

Others preferred to plant crops. Their efforts were aided by the Long Island Food Reserve Battalion, a group of private citizens including William K. Vanderbilt, J. P. Morgan, and Theodore Roosevelt, who offered agricultural equipment and advice to anyone who was willing to plant crops to aid the war effort. Once the crops were raised, Ralph Peters arranged for a specially equipped Long Island Rail Road train to cross the island distributing information about canning.

The American Automobile Association and the Long Island Automobile Club made a different kind of contribution to the war effort by providing free transportation over Mr. Vanderbilt's motor parkway to Camp Upton for relatives of soldiers. One wonders how many of those taking advantage of the free transportation realized that the motor parkway had started out as a race track for the wealthy; but after all, that was more than a dozen years ago, and many things had changed since the days of the Vanderbilt Cup races.

The automobile was no longer regarded as a rich man's toy. Indeed, newer and faster autos along with improved railroads were bringing many nonmillionaires to settle along the Sound in burgeoning suburban communities. At first this phenomenon was limited to the North Shore of Long Island, but by the end of World War I it was affecting the mainland side of the Sound as well. The Sunday *Times* real estate section could thus note by 1918:[11]

. . . the principal buying has been in Westchester and the section adjacent to the North Shore of Long Island Sound in Connecticut. These two latter districts have been favored not only because of the greater supply of land available for country homes, but because of the greater variety of the land and its excellent connection with the city both by rail and automobile roads.

In the second decade of the twentieth century, Daniel Webster's "American Mediterranean" was on its way to becoming a high-class Levittown!

J. P. Morgan Estate on Morgan's Island (Photo by Allan M. Eddy Jr.)

The Great Gatsby and the Roaring Twenties

THE END OF WORLD WAR I OPENED A NEW ERA IN AMERICAN SOCIAL AND ECONOMIC history often referred to as the age of the automobile. With peace restored and the expansive forces of the nation liberated, the impact of the car became no less than revolutionary. In 1915 there were about 2.5 million cars in the country. This number jumped to over 9 million in 1920 and shot up to nearly 20 million in 1925. The sudden arrival of all these power-driven vehicles, invading the traditional ways and thinking which still lingered from the nineteenth century, resulted in the upsetting of standards, morals, and habits. This created, among other things, a rapid expansion of the city *outward*. Long Island Sound, being part of the largest metropolitan area in the country, would feel the impact of this explosion, the immediate effect of which was to change the Sound's appearance and everyday life radically. In a nutshell, the essence of the automobile revolution was that a technological miracle put the steering wheel of the car, formerly the toy of the rich, into the hands of the common man, with innumerable consequences, and nowhere were these felt more than on Long Island's North Shore, *Great Gatsby* country.

F. Scott Fitzgerald on Long Island

One of the principal characters in F. Scott Fitzgerald's celebrated novel *The Great Gatsby* makes the observation that "young men didn't . . . drift coolly out of nowhere and buy a palace on Long Island Sound."[1] Nick Carraway, the character who utters the statement, is referring to his next door neighbor Gatsby, a man with a mysterious past who seemingly drifted in out of nowhere and not only bought a palace on the Sound but peopled it on summer weekends with a dazzling array of sophisticated guests who came from near and far to attend fabulous parties. F. Scott Fitzgerald, creator of the character named Gatsby, ought to have known a great deal about the world described in the novel, because for a time he lived in the midst of it on Long Island's Gold Coast.

The Fitzgeralds—the author, his wife Zelda, and their daughter Scottie—came in 1922 from the Middle West to Great Neck, where they rented a mansion for $300 a month. They were going to live on a yearly budget of $36,000. Handsome as this sum was in the early twenties, it would not cover the expenses of high living, constant traveling to and from New York, and an open house where one party with plenty of antics and long revelries—Prohibition notwithstanding—would follow the other. The newcomer to the Gold Coast quickly found out that $50,000 was the starting line to keep in style among neighbors like the great Gatsby who could throw bigger and noisier parties. Great Neck, as the author described it, was a boom town invaded by the nouveau riche, who grabbed up expensive mansions as fast as they could be built. When he arrived, there were only seven merchants in town, but the boom brought an invasion of money-hungry grocerymen who borrowed heavily from the banks to start their businesses; the only way they could pay back their debts was to raise prices. Great Neck, in consequence, became a very expensive place, perhaps the most expensive in the world, the novelist wrote, evidently not without some poetic license.

How much truth is contained in some of the statements made by Fitzgerald in *The Great Gatsby* is still a matter of dispute. Undeniably the Gold Coast was declining in the 1920s. Like other areas, it began to feel the effects of suburban sprawl. Slowly the fabulous way of life for which the North Shore was so famous in the years before World War I had to give way. But it was by no means over. Anyone who doubted this just had to pick up his morning newspaper and read about the social life on the North Shore.

The Visit of the Prince of Wales

Probably the most exciting social event of the entire decade was the visit of the Prince of Wales in 1924. In order to celebrate the occasion properly, Clarence Mackay gave a fabulous party at his Harbor Hill estate overlooking the Sound at Roslyn. Fashioned after the Maison Lafitte in France and twice as large as the principality of Monaco, Mackay's estate was constructed by seven hundred artisans. The main house boasted a ballroom two stories high adorned with an enormous cut-glass chandelier. Needless to say, Clarence Mackay, heir to a huge fortune was able to entertain his princely guest in pomp and style without making much of a dent in his millions. With that much money he did not have to skimp on the preparations for the gala. Far from it! A special addition to the house was built to accommodate the overflow guests. The mile-long driveway leading to the fifty-room mansion was lighted by hundreds of tiny blue bulbs. Thousands of colored lanterns were strung through the woods of the estate, while floodlights illuminated the mansion. No expense seems to have been spared when it came to refreshments, decorations, or entertainment. The finest foods and wines were served to the twelve hundred guests at the late night buffet which was the highlight of the festivities. The service on the richly decorated tables at the dinner preceding the gala was made by Tiffany and Company from silver taken from the Nevada mines of the host's father John W. Mackay. After dinner the guests danced to the melodies of the Paul Whiteman orchestra on the covered terrace overlooking the Sound. Impressed by the importance of the occasion or perhaps by its symbolic meaning, members of Mackay's staff made notes which have been preserved in the Bryant Room of the Roslyn Public Library. This material is a unique source of information about the glitter and tinsel of the Gatsbyish era on the Gold Coast. But much more was to come!

Lindbergh, Movie Greats, and the Mere Rich

In 1927 Clarence Mackay hosted another lavish party. This time he was honoring a different kind of celebrity, an American hero of the twenties, Charles Lindbergh. Once again the grounds of the estate were bathed in the pale light of red, white, and blue lanterns. The fountains were ablaze with lights, and the house came to life with five hundred guests who assembled to honor the shy young man who had made aviation history by flying solo across the Atlantic. So great was Lindbergh's popularity that, in order to prevent him from being crushed by the crowds, an escort of one hundred New York City motorcycle policemen accompanied him on his way to the Mackay estate, where he was entertained at a formal dinner. After dinner a reception for Mackay's friends was to follow, but the shy young aviator, tired of the endless round of festivities, slipped out and went back to New York. The startled host, informed about the unceremonious departure of the guest of honor, was speechless. Then, recovering from the shock, he ordered the orchestra to keep playing and the servants to keep the champagne flowing. Those attending the reception were given the impression that the guest of honor would join them shortly after a confidential tête-à-tête with the host. But actually Mackay had gone to bed, and the guests were told around 1 A.M. that Lindbergh had left just a short while ago.

Lindbergh's achievement is regarded by some not only as a milestone in transportation but as a bold step which reestablished the nation's faith in the ability of a nice, quiet, ordinary boy to rise to the pinnacle of success. In the late 1920s when Lindbergh flew across the Atlantic, Americans, say some social observers, were becoming disenchanted with bootleggers, gangsters, and other celebrities created by the roaring twenties. Movie stars, on the other hand, were still quite popular in the Sound shore communities where some of the famous silent films were made. In Mamaroneck, for example, the former estate of Henry M. Flagler became the D. W. Griffith movie studios. The story is told of how Mr. Griffith, while residing on the North Shore of Long Island, spotted the piece of land called Satan's Toe jutting out into the Sound at Mamaroneck. He liked what he saw so much that after closer inspection he bought the property, transforming the beautiful mansion, much to the dismay of area residents, into dressing rooms and wardrobe areas. Then he proceeded to build a large structure to house the other aspects of his movie business. A new type of enterprise sprang up, employing many local people as well as school-age youngsters who worked as extras in his movies during summer vacation. D. W. Griffith and other movie moguls soon found out that the Sound and the picturesque shoreline with its old towns could serve as excellent background scenery for their films. Scenes for *The Perils of Pauline* and *When Knighthood Was in Flower* were shot in Greenwich, and *Huckleberry Finn* on Hunter Island.

Although stories about the Barrymores, who had a summer place in Larchmont, and Mary Pickford, who was courted by Douglas Fairbanks, Sr., in Larchmont Manor, charmed everyone, by the late 1920s it was apparent that Americans were bored with the make believe world of the film. They longed for real heroes, that genuine American type, the local boy who by his own efforts rose to national fame. Perhaps even the Gold Coast millionaires were becoming a bit disenchanted with the big parties and the insane antics such as jumping into the pool with their clothes on or, by way of counterpoint, jumping in after underwear and lingerie were flung into the bushes. Coming to an end was life as F. Scott Fitzgerald described it in *The Great Gatsby*:[2]

There was music from my neighbor's house through the summer night. In his blue gardens men and girls came and went like moths among the whisperings and the champagne and the stars. At high tide in the afternoon I watched his guests diving from the tower of his raft, or taking the sun on the hot sand of his beach while his two boats slit the waters of the Sound, drawing aquaplanes over cataracts of foam. . . . On Mondays eight servants, including an extra gardner, toiled all day with mops and scrubbing-brushes and hammers and garden-shears, repairing the ravages of the night before.

The Rum Runners

So great was the rush of the nouveau riche to the Gold Coast, which like a Park Avenue address had the magic quality of bestowing by the mere sound of its name the semblance if not the essence of distinction, that by the end of the twenties the North Shore was dotted with mansions of all sizes and styles. When festive occasions

Aerial view of the Otto Kahn estate in Cold Spring Harbor (Nassau County Historical Museum)

Southport Harbor, Connecticut (Fairfield Historical Society)

The Spirit of St. Louis (Nassau County Historical Museum)

Charles A. Lindbergh at Roosevelt Field, 1927 (Nassau County Historical Museum)

or a simple weekend with guests approached, the problem during the era of Prohibition was not where to celebrate but what to celebrate with. When there is great demand, however, the supply will usually take care of itself in one way or another, and there were many ways on Long Island Sound with its innumerable inlets and harbors, islands and coves, where fast motor boats could slide in and out carrying the precious but forbidden cargo of the rumrunners. The North Shore became their favorite haunt, which in a way continued the tradition of pirates, spies, and smugglers prevalent on the Sound in bygone days.

Some of the estates were excellent places for plying the rum-running trade, especially when the owner was away and the mice could frolic in the absence of the cat. At the Sands Point estate of Solomon Guggenheim, rumrunners were caught by the police, who confiscated $90,000 worth of "imports." When the authorities saw trucks coming from the estate, they became suspicious, since the owner of the place had died the previous year and the estate was for sale. Driving to the dock, the lawmen found a dozen rumrunners unloading cases of Chinese whiskey, a brew described as "the TNT of booze."[3] It was reputed to deliver more kicks per sip than any other hard liquor. (Not much was known about this powerful drink in those days, but a later generation would watch President Nixon and Prime Minister Chou En-lai taking microscopic sips of this liquid TNT from miniature glasses as they toasted each other and their respective countries during the American president's visit to Peking in 1972.) The raids on the Guggenheim estate and many other places were referred to as the Battle of Long Island, that is, the roaring twenties version, a battle waged not against redcoats but against bootleggers and rumrunners.

But it was not always warfare, even though local papers proclaimed ever-new victories. Sometimes the population demonstrated a lack of zeal in cooperating with the authorities. This is shown by an incident which occurred when a ship carrying liquor ran onto the beach at Bayville during one of those fierce storms which can sometimes engulf the Sound. Before notifying the federal authorities, some of the local citizens made a beeline for the beach and helped themselves to the forbidden stuff. Such episodes led to complaints about leaks in the Long Island liquor blockade, but, as was well known to anyone who cared, leaks existed all over. The situation on Long Island was so bad, however, that the Treasury Department in waging war against the smugglers welcomed the assistance of private citizens, some of whom apparently wanted to drive out the bootleggers. Such were the members of the Ku Klux Klan, or the citizens of Oyster Bay who wrote to the district attorney of Nassau County demanding that five saloons or "hell holes" in their town be closed. With outrage the latter noted that such places were allowed to exist on one of the principal streets. Worse still, said the irate citizens, visitors to the grave of Theodore Roosevelt had to pass the saloons. All in vain! For no amount of complaint or indignation could keep the profiteers of prohibition from plying their trade and garnering their profits. Like narcotics pushers of a later age, they were indifferent to human suffering or degradation. They were ruthless but primitive gamblers who thought that by applying violence they could bend fortune to their will. The gamble often failed; if caught, they invariably suffered the loss of their goods as well as the loss of personal freedom since a jail sentence usually awaited them. Sometimes, however, the punishment was a bit more unusual, such as the sentence given a Bridgeport man who had transported illegal beer. He had to open 6,624 bottles and pour their contents into a sewer. The Bridgeport man's fate was probably no worse than that of a New Haven woman whose husband paid his own fine for selling liquor but allowed his wife to go to jail. New Haven had other problems, too. Liquor was finding its way into the veterans' hospital and onto the Yale University campus where the so-called varsity bootlegger was said to be bringing liquor directly to the students' rooms.

The End of the Golden Age

No history of the North Shore would be complete without mentioning Sagamore Hill, the classic monument of an age when the prominent, the powerful, the famous, or the merely rich erected palaces on Long Island. Many of these monuments have since been turned into tourist attractions, museums, colleges, or, as is the case with Sagamore Hill, national shrines. Falaise, the Norman castle built by Harry Guggenheim at Sands Point, is now part of the Nassau County Museum, and the William Vanderbilt estate at Centerport is owned by Suffolk County, which opens it to the public during the summer.

High society of every historic period will select a geographic area where the grandees will congregate to fill their abundant leisure time with conspicuous consumption in some form, depending upon the tastes and styles of the age. But whether it is in the valley of the Loire, Newport, or the North Shore of Long Island, the social behavior will express the character of the elite who inhabit the ornate structures and play games and sports in the surrounding gardens. The Gold Coast symbolized the crest of the wave which brought America to the forefront of events as the richest and most powerful country in the world, especially after World War I. That crest was solid before the war, the triumphant top of the wave created by the rapid business expansion at the turn of the century. It became frothy in the tumult of the roaring twenties, and finally it turned into shreds of mist as the storms of the Great Depression swept over it.

106

The Real Estate Boom on Long Island

The seemingly abundant supplies of liquor on both sides of the Sound may have caused problems during the 1920s; nevertheless, many residents of the Sound shore communities were little concerned with bootlegging and rumrunning. Instead, they were absorbed in living the good life, popularly expected to be the reward of those who dwelt along the Sound. Some Sound shore areas, particularly on Long Island, were experiencing a big influx of population, or, as one *New York Times* writer put it:[4]

Boom times have come to Long Island. . . . Long known in fiction as the abode of English butlers and American millionaires, suspected of undemocratic snobbery, vaguely visualized as a piece of land bounded on the north by the Prince of Wales and on the south by an an ocean of rum runners . . .

This may have been exaggerated a bit, but to many people who sought homes in the suburban developments, the North Shore appeared as an idealized image of Long Island which they wanted to retain. Although the Gold Coast was declining, some of its new and less affluent residents wanted to do things on a grand scale. A group of architects, artists, and sculptors even made a proposal to erect large multistory chateaux overlooking Manhasset Bay. The houses in this "New Versailles" would each be on two hundred acres dotted with fountains and sculpture. It was also in Manhasset that the Metropolitan Museum of Art sponsored the development of a community called Munsey Park, named after Frank Munsey, who had left three hundred acres of his estate to the museum. Farther west at Port Washington the Bourke Cochran estate, which had an excellent view of the Sound, was sold to a developer. Similar sales followed everywhere in Nassau County. The breakup of the large estates occurred with such speed that some of the North Shore communities decided to adopt master plans, the aim of which, as the Great Neck plan stated, was "to prevent spoiling of pleasant residence districts by uncontrolled land speculation and reckless building."[5] Planning seemed diametrically opposed to the big land boom which hit Long Island in the mid-twenties. It appeared somewhat incompatible with the Long Island Real Estate Board's slogan "Long Island: Isle of Opportunity."[6] On the other hand, by the mid-1920s it was evident that development could not proceed unchecked indefinitely. Parts of Nassau County close to New York City were already built up, and further development seemed neither warranted nor desirable in the opinion of the older residents.

The Growth of Westchester

On the other side of the Sound, it was a similar situation, except that on the mainland some sections were more liberal in opening their area for development than the exclusive North Shore communities. In the Throgs Neck region of the Bronx, for instance, former estates, including that of Collis P. Huntington, became homesites for thousands of new dwellings. Unlike most of the North Shore of Long Island and the more restricted areas of Westchester County to the north, Throgs Neck in 1895 became part of the great metropolis of New York, thereby quickly losing its rural character. An elevated subway connected the area directly to Manhattan. Before the invasion could engulf them, the millionaires fled, selling their estates to the developers.

Some of the wealthy refugees from Throgs Neck fled to the North Shore of Long Island, but Westchester County was equally appealing except for the fact that even there the masses were making inroads. One writer summed the whole thing up by stating: "Westchester County is the natural outlet for New York, free from ferries and bridges. . . . The county is fortunate in being bounded on the east by Long Island Sound."[7] That the Sound attracted numerous exurbanites to Westchester is evidenced by the growth of the communities located on the waterway. Immediately after World War I a certain Sound shore village, Pelham, was described as being on the verge of the greatest boom in Westchester history. Encouraged by this rapid expansion, one Westchester realtor spoke in glowing terms about the future of the county, declaring in 1924 that "the only thing which could possibly upset us would be an earthquake or a war, and it is the extreme pessimist who keeps his eye open for such calamities."[8] On a spring day two years later, a tremor did hit the eastern part of the county. It was felt in the Sound shore communities from New Rochelle to Port Chester, where it shook houses and rattled dishes. Startled suburbanites in New Rochelle, Larchmont, Mamaroneck, Rye, and Port Chester reported the vibrations which the seismograph at Fordham University recorded as an earthquake.

What Is a Home Without a Nearby Golf Course?

Reports of an earthquake in normally stable Westchester County were not enough to deter suburban homeseekers from migrating to the pleasant area north of New York City. In fact, during the remaining years of the 1920s Westchester was the preferred location for New Yorkers. Some of the most desirable property in the county was found on the Sound shore wherever estates near the water were being subdivided for building lots. For the less affluent who could not afford even a smaller home on the waterfront, a moderately priced house in a Sound shore community was the answer. In some towns and villages along the Sound, however, such homes were difficult to find, especially in the southern part of Fairfield County. Although demand for houses rose greatly, some of the communities like Greenwich simply did not offer any bargains in real estate. Greenwich was a mecca of wealth

108

attracting the kind of people who worry more about their investments and their yachts than about paying off the mortgage on a little cottage. Occasionally an event might break the monotony of their good life, as when one summer morning in 1928 two deer mounted the steps of the Greenwich Country Club and wandered into the lounge and out onto the veranda where startled members were having breakfast. But generally the concerns of the Greenwich citizenry were not so mundane. The same holds true for members of the Westchester Country Club, the fabulous new establishment developed in the 1920s at Harrison. Hidden inland among the trees, the main clubhouse, really a multistory hotel, offered a glorious view of the Sound from the upper floors. To accommodate its members, the club maintained a private beach at Manursing Island in Rye. Perhaps its biggest attraction was the superb golf course. In the 1920s proximity to a golf course was an important consideration for a homesite; as one Westchester County realtor put it, "What is a home without a nearby golf course?"[9] There may have been something to this statement, considering the great interest in sports of all types displayed by the more leisurely classes. But what about the new wave of suburbanites who flocked northward from New York City during the 1920s? They could afford to purchase a home, but they could hardly go beyond that. Membership in a private club being out of the question, they sought recreation in the public parks and beaches.

Parks and Parkways in Westchester

In this respect a new homeowner was probably better off in Westchester than anywhere else, because the county had a growing network of scenic parks and recreational areas available to residents at a very nominal cost. As early as 1924 Westchester County with its own park commission could proudly claim to have the biggest park and parkway system in the state. Much of the credit for this achievement goes to the far-sighted Westchester County Board of Supervisors' appropriating funds for the acquisition of open spaces as soon as they became available. In 1921, for example, the supervisors conceived the idea of transforming Rye Beach Amusement Park, a privately operated Coney Island type place, into a county park with a stretch of waterfront regarded at the time as one of the best beaches on the Sound. In 1926 when nearby Paradise Park burned down, the park commission decided to clear the whole area for a modern mass facility. Initially the residents of Rye doubted whether county management would bring an improvement over the riotous conditions of the old amusement park. But on July 4 1928 when the new county park opened, the park commission lived up to its original pledge "to operate the amusement park under proper restrictions and to eliminate rowdyism and petting parties."[10]

The same high standards were in force at Glen Island, Westchester County's other Sound-side park in New Rochelle. By 1923 when the county purchased the once famous resort from John Starin of New Haven, the buildings had deteriorated to the point where the county decided to demolish them, leaving just open spaces and the beach. The county also built a bridge to the island to replace the old ferry, the only way of access to Starin's Amusement Park. There was considerable opposition among New Rochelle residents to the new project, because land had to be acquired for the approach roads to Glen Island. Some New Rochellians undoubtedly thought that the defunct plan to acquire the island as a war memorial should be revived. What really bothered the local people was the possibility of a wholesale inundation of the new facility by residents of the nearby Bronx. Such a development could easily have been avoided in the nineteenth century when most of the visitors to Glen Island came by Sound steamer or the New Haven Railroad. Permission to land could have been denied the New York City boats. In the 1920s, however, things were not that simple. Most of the visitors arrived by car, and, short of issuing residence permits, Westchester County could do little to keep the neighboring Bronxites out.

In some ways Westchester was actually encouraging residents of New York City to inundate the county. For example, Westchester built the first parkway in the United States, a road paralleling the Bronx River. This new highway brought huge crowds of New Yorkers into the county, but even before its completion traffic on the older routes had increased to the point where relief became imperative. According to estimates made in 1925, fifty thousand cars per day used the Boston Post Road in Westchester County alone. Under such traffic loads pavements wore out quickly. Constant repairs slowed down the movement of vehicles, which did not improve air quality. Yet the local population, desperate about the unending din and heavy fumes, opposed projects to widen the road. That, they said with prophetic foresight, would only bring more cars and trucks. Pelham Manor home-owners actually staged demonstrations to prevent work on the road.

The Westchester County Board of Supervisors, confronted with a determined opposition to widening the Boston Post Road yet keenly aware of its steadily deteriorating condition, seriously considered building a new highway running through the Sound shore communities. Such a measure was actually proposed in 1925, and by 1929 the county parks commission in charge of these matters had acquired all the necessary land to build a new road from Pelham to Port Chester, spanning the entire length of the county along the Sound. Then came the Great Depression, bringing in its wake empty treasuries in Westchester as well as in Albany, and the ambitious plan had to be postponed till better times. World War II lengthened the postponement by many more years so that the project could not be resumed before the fifties. Meanwhile the Hutchinson River Parkway helped to relieve the congestion on the narrow streets of the Sound shore communities. The

110

"Hutch," winding its way through beautiful scenery, quickly became, like the Bronx River Parkway on which it was modeled, a very popular motor road. Built in the age of thirty-five-miles-per-hour limits, it afforded the driver a thoroughly enjoyable weekend spin on its pleasantly landscaped curves. Today, in the age of the sixty plus (mostly plus until the energy crisis) speeds, curves in the roadway, no matter how scenic, are an annoying and dangerous impediment to the hurried motorist.

Westchester actually acquired world fame by adjusting so quickly and so successfully to the demands of the onrushing automobile age. In 1928 it had 140 miles of parkways and had spent or appropriated $47 million on roads. Experts came to study this system which made living more attractive and healthy by eliminating swamps and salt marshes, converting them into parks, beaches, or roads. The swelling travelers were thus channeled into scenic routes leading to amusement centers, forest preserves, or playgrounds. No wonder the excellent communication on the parkway led to a rapid increase of property values, which, on the whole, doubled in the decade of the twenties. Along the Bronx River Parkway property values within a zone of five hundred feet on either side of the road rose 800 percent! It is thus no surprise that landowners were generous in cooperating with the authorities. But in its sixth annual report to the Board of Supervisors, the Westchester County Park Commission listed many land donations, the largest of which, amounting to fifty-six acres, came from John D. Rockefeller.

Long Island Lagging Behind

On the other side of the Sound, highway building in the 1920s had a similar effect on real estate development. Initially, though, the coming of good roads to Long Island failed to produce a rise in real estate values on the North Shore. To understand this problem, it is necessary to go back to the period immediately following World War I. When the highway system of the country, built in the days of the horse drawn wagon and then neglected because of the rapid spread of the railroad network, suddenly became inundated by a new type of vehicle, an emergency arose for which the authorities had no ready-made solution. Driven thirty to forty miles an hour instead of ten, this new vehicle stirred up clouds of dust, was noisy, and emitted noxious fumes. It was also dangerous to pedestrians and other vehicles and even more so to its own driver. In one way the automobile became very popular and was bought eagerly by all who could afford it—and Henry Ford made that possible for millions—but in another way, since there were few parks or playgrounds inside the city limits, the new mode of travel produced a destructive exodus, especially on weekends, from the metropolitan area to the beaches and woods of Long Island. To the people who lived near these beaches and woods, used

111

to the peaceful quiet and serene surroundings of the country, the invasion of the horseless carriage bringing masses of picnickers appeared highly undesirable. They reacted with bitterness and hostility, but to no avail. The crowds increased every year, pushing their way deeper and deeper into woods, meadows, even gardens and orchards, or to the seashore, leaving behind litter and trampled-down vegetation. Signs were ignored, fences broken down, and private property abused. Inhabitants of Long Island, "the natural playground of New York City,"[11] looked forward to every weekend in summer with horror and dismay. "The tides that swept upon King Canute were no more inexorable,"[12] commented a *New York Times* editorial early in 1926.

Enter Robert Moses

One solution to this gigantic problem was to follow the example of Westchester County and create a network of parks which would funnel the crowds away from private property. This was the idea proposed by Robert Moses, whose life's aim was to serve the public by creating parks, playgrounds, and recreational facilities wherever there was room, opportunity, or urgent need to do so. He managed, not without difficulty, to build a most magnificent beach facility on the South Shore, together with a large park. The fame of Jones Beach spread the world over, and Heckscher Park is one of the largest recreational areas in the state. To reach them, Robert Moses devised a network of parkways leading from Manhattan, Brooklyn, Queens, and eventually also from the Bronx to the seashore. His plans stirred up much adverse comment and active opposition. Many alternative solutions were offered, but the urgent thing was to do something, for as Moses himself once said, there may be many men with bright ideas but very few who can carry them out. He was a rare species, and his determination to go ahead was not hampered by excessive modesty. Approaches to Jones Beach and Heckscher Park were soon completed by a magnificent system of wide highways and parkways, but on the northern part of the island he ran into heavy obstacles which were not easily overcome.

According to the plans of the Long Island State Park Commission, a parkway was to be built in the northern part of Nassau and Suffolk counties, cutting through the estates in a straight easterly direction. On it would flow the enormous traffic leading from the metropolitan area to beaches and parks which had been built or were in the course of construction in the northern as well as the southern parts of the island. To prevent the intrusion of the masses into the treasured preserves of wealth, various interest groups turned up among the estate owners to defeat the project or, as Governor Alfred Smith put it in his annual message to the legislature, "to keep and preserve the wall that had stood between the millions of the city and

112

their luxurious estates."[13] These people had powerful connections and influence. For instance, Senator George L. Thompson, representing Nassau and Suffolk counties in Albany, proposed in 1926 to abolish the Long Island State Park Commission because only this authority could spend money appropriated from the bond issue that was to finance the construction of parks and parkways. But Governor Smith, a staunch friend of the people, sided with Moses and held the fort by vetoing bills passed by the legislature favoring Long Island's elite group. When one complained that their exclusive domain would be overrun by "the rabble," Governor Smith, a native of the lower East Side of Manhattan, exclaimed that he was part of that rabble, and vetoed again. Another incident occurred when one of the matrons who loved fox hunting complained that the proposed parkway would confine the fields to a small area, so that if the fox crossed the parkway, the dogs would lose the scent and could not pursue their quarry. Facetiously Moses suggested that viaducts built under the road might provide passage for foxes and dogs alike.

The Battle of Wheatley Hills

A most serious objection to the parkway arose when a group of estate owners in the Wheatley Hills section of Old Westbury, their property line astride the right of way as planned by the park commission, formed a group in 1929 to agitate for a new plan. They suggested that the old Vanderbilt Motor Road built in 1904 should be taken over by the state and developed into a modern parkway. Moses immediately went to the counterattack, pointing out that this was a stratagem of the millionaire landowners to divert the route of the Northern State Parkway which was to cross their estates. In responding to the charges of Moses, the members of this group, called the Nassau County Citizens Committee, met at the Harvard Club in New York, declaring in their resolution that Moses misrepresented them and their intentions. They claimed that they were not just a handful of very wealthy people who were playing a selfish game of obstruction. On the contrary, they listed 264 names as active members of the group representing owners of 18,000 acres, which is 4,000 more than the whole of Manhattan Island. Their intention was to save money for the state by offering the plan to purchase the Motor Road instead of building a new parkway, which would be more expensive.

Moses summed up his case and counterarguments indirectly. In reply to a letter asking him why there was such a long delay in building a parkway in the northern half of Long Island, Moses reviewed the fight he had had to wage against the estate owners since 1924, that is, since the Long Island State Park Commission began its work. Opposition to the commission was always intense and continuous, he said. There were vigilante committees in Albany doing their best to prevent the parkway system from becoming a reality. Their latest device to divert attention to the Motor Road deserved no serious consideration because, to begin with, the road was poorly constructed and its location, moreover, was in the middle of the island and not in the northern part where a modern parkway was badly needed. He also complained about reluctance of public officials in these northern sections of the island to give support to the parkway plans. They sided with the estate owners. This is evident from the underassessment of the large estates, or their listing on the tax rolls as being much smaller than their actual acreage, so that the taxes paid were small in proportion to the size of the holdings. The opposition of the estate owners to the plans of the park commission went to the length of trying to defeat appropriations in Albany for surveys. Yet it was evident that, because of the rapid development of the parkway and highway program in Queens, traffic conditions in the northern half of Nassau would, in a few years, become desperate. But the estate owners hoped that they could force this traffic down to the south side of the island, where there were several arterial roads. Moses, however, knowing the attitude of New York motorists, assured his correspondent that the traffic would not go south. It would try to seek ways and means to continue east, clogging the narrow roads, causing giant traffic jams and tragic accidents. He finished his letter by suggesting that the quarrel of this correspondent should not be with the State Park Commission but with the Nassau County Citizens Committee.

The Bend in the Parkway

The opposition of the estate owners, however, was too powerful to be broken. Only a compromise was possible. It took place in the office of ex-Governor Alfred E. Smith in the City of New York toward the end of 1929. The right of way of the Northern State Parkway had to be changed so that over a distance of 2.3 miles it would go south to avoid crossing the Wheatley Hills area. New land had to be acquired by purchase, and the estate owners were willing to contribute $175,000 for that purpose. This stretch is now an eight-lane parkway accommodating an enormous flow of traffic that is funneled into it in its northern elbow from the Long Island Expressway and into its southern elbow from the Meadowbrook Parkway. The Wheatley Hills section of Old Westbury and adjoining communities is still beautiful, unspoiled countryside dotted with old mansions and palatial residences, many of which in the meantime have been taken over by colleges and universities.

The compromise between the State Park Commission and the Nassau County Citizens Committee removed the worst obstacle in the way of the Northern State Parkway. Its construction went ahead somewhat slowed down by the Great Depression, but soon, with transverse connections leading to the Southern State Parkway, a magnificent network of landscaped four-lane roads emerged on Long Island, on which the motorist, kept within moderate speed limits, could enjoy the countryside on his way to the parks and beaches of the island.

114

The Motor Age Invades Connecticut

If Long Island lagged behind Westchester in adjusting to the demands of the automobile, Connecticut fell even further behind. The state highway department, seeking a radical solution, decided to build a highway from Greenwich to New Haven paralleling the Boston Post Road, but, as in Westchester, the proposal was rejected. Meanwhile truck traffic increased rapidly, especially short-haul traffic, which was simpler, cheaper, and faster than rail. While highways, as a consequence, became an important economic asset, they were not without heavy drawbacks. Truck traffic with its dust, dirt, fumes, and noise became a big nuisance, its heavy rumble shaking the old buildings facing the narrow streets of Connecticut's towns on the Boston Post Road. These towns, as in Westchester, follow each other at short distances, forming already in the twenties an almost continuous string of built-up areas. Since trucks at this time were especially suited for short-haul traffic, they quickly replaced the railroad in this densely populated industrial region.

The number of passenger cars increased with equal rapidity, and in the late twenties they invaded the quiet interior of the state, bringing with them the blessings and woes of the automobile age. The great attraction of rural Connecticut is its charming colonial villages with their broad streets, venerable elms, and carefully cultivated lawns and gardens. Like a living picture book these colonial towns and villages unfold before the traveler passing through them in the fast-moving car, recalling the quiet, simple, good old days, that offered such an attractive contrast with the noisy, hurried, rough-and-tumble present. This pleasant world, so dear to natives and visitors alike, was now threatened by the onrush of the new age bringing hordes of tourists demanding more room and accommodations and services. The signs of change were springing up all over the peaceful countryside, defacing the roads with billboards, gas stations, and hamburger stands.

But the worst consequence of the increase of traffic was the necessity of widening the streets to make more lanes available for cars and trucks. In many cases rows of old shade trees had to be destroyed, raising a great outcry among the natives. Connecticut, they said, had not only been industrialized but now materialized, and even foreignized by the legion of out-of-state visitors coming in their cars to enjoy the landscape and charming villages of the state where, in some cases, parts of old homes had to be either cut off or moved. The state's Forest Association asked for legislative measures to prevent further despoiling of treasures which would be irreplaceable, and the State Park and Forest Commission decided to build parkways like those in Westchester which avoided the towns and villages. It was indeed to be an early salutary measure of conservation which at the same time modernized the transportation of the state, at least as far as passenger traffic was concerned.

Auld Lang Syne

Thus, the motor age spread all around the Sound by the end of the twenties, causing battles between local interests and the needs of the wider public on Long Island, upsetting the traditional ways of Connecticut, but dealt with very efficiently by Westchester. Everywhere the solution, whatever it happened to be, was only another stage in the relentless development of the Sound area, for problems in the modern age can never be solved permanently. Onrushing events represented by more cars, more needs and demands, would make any solution only temporary, so that it is almost proverbial now, as we enter the last quarter of the twentieth century, that any new highway as soon as it is opened proves to be inadequate to handle the traffic.

The compromise on Long Island creating the bend in the Northern State Parkway was no exception. Traffic soon demanded additional measures which still respected the quiet of the parklike expanse of the Wheatley Hills. It was only slightly disturbed by the thousands of cars and trucks rolling along the six-lane Long Island Expressway which, in a straight line, now bisected the southern portion of this area. This modern artery does not change the region's general appearance of serene beauty, because it runs in a deep cut below the level of the other roads in the vicinity. Local traffic is light, and the expressway is not visible unless the motorist crosses it on one of the many bridges. It is said that when Robert Moses was tracing the right of way of the Long Island Expressway and the estate owners again came with their objections, he simply waved them aside by saying that you cannot accommodate expressway traffic with sharp bends in the road, nor would public opinion bend to the wishes of a handful of wealthy men. The only concession made to the millionaires of Westbury was that the portion of the Long Island Expressway which runs through their area was placed in a cut, but this was not much of a victory compared with their 1929 triumph.

Long Island had changed a great deal since the 1920s, however. No longer was it the preserve of the rich. Now it belonged to the masses. The estate owners, looking back on the rapid development of the island since the building of the Northern State Parkway, could perhaps commiserate with Nick Carraway in *The Great Gatsby,* who described the end of the summer season as follows:[14]

Most of the big shore places were closed now and there were hardly any lights except the shadowy, moving glow of a ferry boat across the Sound. And as the moon rose higher the inessential houses began to melt away until gradually I became aware of the old island here that flowered once for Dutch sailors' eyes—a fresh, green breast of the new world. Its vanished trees, the trees that had made way for Gatsby's house, had once pondered in whispers to the last and greatest of all human dreams; for a transitory enchanted moment man must have held his breath in the presence of this continent, compelled into an aesthetic contemplation he neither understood nor desired, face to

116

face for the last time in history with something commensurate to his capacity for wonder.

Whether Nick Carraway realized it or not, the Long Island of the early twenties still resembled conditions known to the Dutch more than it ever would in the future. Indeed, the Gatsbyish era was a turning point not only in the island's history but in that of all the Sound shore communities. With the steering wheel of his automobile in his hand, the common man drove even farther from the big city, encroaching upon the preserve of the rich, invading the quiet old towns and villages of the early colonists, demanding more highways and playgrounds. and forever foreignizing, materializing, and suburbanizing. Yet no one knew at the time that the arguments, fights, and excitement provoked by the onrush of the automobile were but the birth pangs of megalopolis.

Castle Gould built by Howard Gould for his actress wife (Nassau County Historical Museum)

1939 World's Fair at Flushing Bay, New York (Photo by Beauregard Bettancourt)

8

The Bleak Years
1930-1945

Onset of the Great Depression

THE BEGINNING OF THE GREAT DEPRESSION IN 1929 BROUGHT MANY CHANGES TO LONG Island Sound. Fortunes made in the heady days of the twenties were lost in a flash, and the carefree lifestyle of the previous decade disappeared forever. The sudden collapse ruined many individuals and businesses, spreading tragedy throughout the country. But nobody shed tears for the demise of one business that had thrived so well on the Sound: rumrunning. It was not a direct casualty of the Great Depression, except that the economic calamity convinced the public as well as the government that legalization of the manufacture and sale of alcoholic beverages might help to revive business.

Direct effects of the depression hit the industrial centers most severely, among them Bridgeport and New Haven, causing vast unemployment. Unable to cope with economic catastrophe, the authorities could not offer relief. What followed is familiar to all students of this period of American history: foreclosures, evictions, bank failures, destitution, the rise of shantytowns, and long lines of people waiting before the doors of charitable institutions and soup kitchens. Long Island, by contrast, could stand the strain much better. In the thirties most of the area had still preserved its rural character, giving way to the heavy influx of population and its attendant problems only after World War II. The North Shore, of course, changed little, remaining the serene land of millionaires, who contrary to popular belief did not all go under in the stock market crash of 1929. Their sufficiently well-diversified investments permitted many a Gold Coast resident to weather the storm of the depression.

119

Even for the average Long Islander, the depression did not seem terribly ominous, at least initially. The Long Island Chamber of Commerce reported that in 1931 Long Island was less affected than any other area of similar size in the United States. The building of homes and industrial plants continued through the early thirties, while Long Island had fewer bank closings than other areas of comparable population. In 1931 alone almost one hundred new industries were established on the island. The following year, 1932, also witnessed some growth, but the number of unemployed increased. Some Long Island municipalities considered reducing real estate taxes in order to help out unemployed homeowners. In other towns, including several along the Sound, new post offices were planned in an attempt to provide work for the growing number of unemployed.

Across the Sound in Westchester County, the number of people on relief increased dramatically. In response to the emergency created by the depression, the Westchester County board of supervisors voted $200,000 to repair the county's roads and thereby provided employment for some of those who otherwise would be on relief. As an economy move, the supervisors considered the possibility of abolishing the Westchester Park Commission. They did not actually take this drastic step, but wages of park employees were reduced. Trying to save wherever possible, the county turned off the lights on Westchester roads, inviting a quick protest from the American Automobile Association; obviously, the measure proposed to economize at the expense of the safety of travelers.

Economizing and Creating Work

Westchester County sought to retrench and economize even in ways that drew sharp criticism. On Long Island, on the other hand, construction of the great arterial road called the Grand Central Parkway in Queens and the Northern State Parkway in Nassau County went on unabated, involving the expenditure of considerable sums of money and giving employment to many workers. In 1932, when the state advertised for bids for paving portions of the route already laid out, it calculated that there would be enough work to employ 3,000 men for thirty hours a week. A year later Robert Moses complained that the Civil Works Administration had not provided equipment funds and that this would mean laying off 2,500 men. Aside from temporary crises like this, work on the Northern Parkway on the island progressed.

Meanwhile across the Sound in Connecticut the Fairfield County Planning Association presented an extensive program in 1937. It envisioned the following goals for southeastern Connecticut: to provide for the free flow of traffic between New York City and New England while preserving the integrity of Fairfield County, and to beautify the county's roads. Interestingly, as a means of realizing its first

120

objective the association proposed "a coastal express highway parallelling the railroad from the Pelham–Port Chester parkway at the New York state line to the New Haven County border at Stratford."[1] In the interim Westchester County had acquired all the land needed for its Pelham–Port Chester parkway. By establishing the Pelham–Port Chester Authority, the county also attempted to obtain $13.5 million for the construction of the road from the Reconstruction Finance Corporaion. Delays caused the plan for this parkway to be put away until after World War II. Instead, a more important Westchester County road, the arterial Hutchinson River Parkway, became a reality in the 1930s when the Westchester County Board of Supervisors appropriated the money to complete the parkway to the Connecticut border. At the same time Connecticut moved ahead with plans to join the Merritt Parkway to the Hutchinson River Parkway. Residents of Fairfield County, especially those who lived in the Sound shore communities, saw the Merritt Parkway as a method of relieving congestion on the Boston Post Road. The Fairfield County Planning Association envisioned the parkway as a factor contributing to the growth of Connecticut, and particularly the southeastern part of the state, since the Merritt Parkway would provide easy access to New York City. Despite a cost overrun and a grand jury inquiry into the acquisition of the land, the Merritt Parkway was built and brought new people into the Sound shore communities of Connecticut.

The Problem of Exclusion and Isolation

Debates were often raging in some of those communities over the merits of a population increase. The citizens of Greenwich, for example, were divided on the issue. Some opposed the invasion of former New York City residents, while others wished to see their town grow. A Greenwich author writing in 1935 summed up the debate very well:[2]

These are the extremes. Between are the great majority of people who love Greenwich as it is today; who would like nothing better than to see it continue about as it is today, but who realize that the gradual encroachment of a great city can no more be resisted than can the tide of Long Island Sound be turned back.

Throughout most of the 1930s the question of growth was not a major issue in the Sound shore communities. In fact, fine homes literally went begging during the Great Depression. Unable to meet their mortgage and tax payments, many homeowners lost their residences. This scenario, played out all over America, seemed especially dismal along the Sound, where some truly magnificent homes were placed on the market at a fraction of their worth. The famous Otto Kahn estate on the Gold Coast erected at a cost of several million dollars during World War I was sold for $100,000 in 1939. In Connecticut and Westchester waterfront mansions and smaller homes changed hands under similar circumstances. Nevertheless, the desire to uphold the standards kept scrupulously in the past remained strong. The exclusive character of these residential communities rested on them. In Westchester County, for example, despite sagging real estate values, adherence to these standards kept the teeming masses of New York City from invading the preserves of the privileged.

Actually, the Westchester debates in the 1930s went on not so much over who should be allowed to buy homes, since sellers were happy to find prospective buyers even if the latter happened to come from a middling neighborhood in the city, as over who should be permitted to use the county's recreational facilities. Other areas along the Sound permitted everyone to use their parks and beaches. The state of Connecticut was busy improving Hammonasset Beach on the Sound, a facility open to everyone. In Westchester, however, the county board of supervisors wondered whether to maintain the open-door policy at the many parks under the jurisdiction of the Westchester County Park Commission and at the county's two beaches on the Sound, Rye Beach and Glen Island. The board of supervisors in July 1932 expressed indignation at what they termed the New York City invasion of Westchester parks. Three years later the Westchester County Park Commission discussed the possibility of placing Westchester County recreational facilities off limits to anyone but county residents. Since the estimated loss for barring New Yorkers would be $1 million per year, the commission gave the matter serious consideration. Some Westchester County parks were holding their own financially despite the depression, in large part because they attracted outsiders. Playland, for example, made a profit of $300,000 in 1932. In his report for that year, the director of the famous amusement center on Long Island Sound stated that Playland had had a better season than any other facility of similar size and scope in the state. The park commission consequently had second thoughts about barring outsiders, even though Westchester was losing money as a result of many city motorists purchasing Westchester County license plates to enable them to obtain a discount when parking at beaches and parks.

Ultimately the county instituted a resident-card procedure in the 1950s when economic losses would not influence the decision. In the 1930s such a measure could not be considered. Not only would it entail serious economic losses, but experience taught that the device could easily be circumvented, and in fact it failed to work satisfactorily. At Rye Town Park on the Sound, reserved for residents, the cards found their way either by purchase or rental to outsiders.

Moses Has No One to Quarrel With

The question of resident cards, and within its framework the larger question as to who should have access to the Sound, defied solution in the thirties, but not so far as Robert Moses, the champion of the common man, was concerned. As early as 1931 he proposed a system of good roads, parkways if possible, connecting the recreational areas of Westchester and Long Island. He could mobilize public opinion and pull influential strings. At a testimonial dinner in his honor attended by one thousand Long Islanders, former Governor Alfred E. Smith spoke about the great vision of "the dreamer — planner — doer"[3] Robert Moses regarding recreation for the masses. He concluded that state and local commissions would eventually own all the shorefront property. Moses himself mentioned some of the difficulties he had encountered in attempting to transform his dreams into reality. Referring to his quarrel with the estate owners over the route of the Northern State Parkway, Moses said: "Now it would seem there is no one left to quarrel with."[4]

. . . Isolationist Westchester Has

If Robert Moses had no one to fight with, Westchester County did. The completion of the Hutchinson River Parkway not only brought hordes of New Yorkers to the county's parks, including the two on the Sound, but the other excellent roads of Westchester tempted many urbanites to take leisurely drives, especially in the beautiful communities along the Sound. Westchester residents, therefore, began to ask why other people should use for nothing roads the county's citizens had paid for. Although rather distant from Westchester County, the Henry Hudson Parkway and the Triborough Bridge, the latter completed in 1936, were expected to funnel a large volume of traffic into the county, overloading the roads. Tolls on Westchester parkways were proposed, but the county's board of supervisors concluded that it would be difficult to collect tolls because the Westchester parkways had so many access roads and at intervals short enough to enable motorists to beat the toll by getting off the parkway one exit before the toll booth and getting back on the next entrance after.

Although these problems were very real, what it all boiled down to was the fact that Westchester residents, like the citizens of the exclusive Fairfield County community of Greenwich, were divided on the question of growth. Those who favored greater development cited the excellent parkway system as a factor in attracting new people, while those who wanted to keep the area exclusive favored closing the parks and beaches to outsiders and charging everyone for use of the parkways. In this way perhaps the urban masses could be kept out. This struggle, continuing well into the post–World War II period, can be traced directly to the automobile. The mass-produced car made travel to the suburbs feasible, and despite the depression the 1930s saw a great deal of travel in the Sound area, both on the new roads located on the periphery of the waterway and on the Sound itself.

Decline and End of Shipping on the Sound

The decade of the thirties witnessed the last grand procession of gleaming ships from New York to New England via the Sound. The demise of the Sound steamers coming in the late thirties was not to be so much the result of the depression but rather the natural consequence of the increased use of autos and trucks. Although steamboats continued to carry raw materials to the factories of Bridgeport and New Haven, they were increasingly eclipsed by fleets of trucks on the Boston Post Road. More important, travelers bound for New England took to the highways instead of booking passage on the great Sound steamers. The result was the end of a long tradition on the Sound. Already in the early 1930s signs of serious trouble appeared. Both passenger and freight business fell off sharply, and a real war ensued between the New Haven Railroad, owner of most of the Sound steamship lines, and the Colonial Navigation Company, operator of the Colonial Line, after the latter complained to the Interstate Commerce Commission about the New Haven Railroad's monopoly of the Sound steamship service. The Colonial Line based its case on the Panama Canal Act of 1914 forbidding a railroad to operate steamships when boats were in competition with the railroad unless specifically authorized to do so by the ICC. Since both the railroad and its steamboats served the same territory, they were definitely in competition with each other, but that is exactly what the directors of the New Haven had in mind. By controlling transportation both on the Sound and parallel to it, the New Haven would be sure of getting most of the business in the Sound area. As for the few competing steamship lines still existing, a secret report allegedly circulated among the railroad executives stated that the New Haven's steamship lines should be coordinated to eliminate the competition. During the ICC hearings the president of the railroad, J. J. Pelley, denied that the New Haven's ownership of the Sound steamers had eliminated competition in the transportation industry. As proof of this, he pointed to the fact that trucks were taking $27 million a year in business away from the New Haven. According to Mr. Pelley, the railroad's Sound steamers had carried 55 percent more freight in 1916 than they were carrying now. The railroad scored additional points during the hearings by calling in as witnesses representatives of New England business firms who testified that the discontinuation of the New Haven's steamship service would have a detrimental effect upon the economy of their region. Some of the witnesses declared that if their shipments had to go by rail, the delays would be intolerable. The railroad's Sound steamers, they said, provided quick overnight service for cargo. One witness referred to the special difficulties encountered in handling fresh fish and concluded that any drastic changes in the existing steamship service would have a serious effect upon fish shipments to New York.

Ironically, the ICC hearings would not make much difference to the ultimate fate of the Sound steamers. Although the railroad was not compelled by the government to divest itself of its steamship interests, by the mid-thirties the New Haven's steamers were in serious difficulty. In 1935 the New London line was discontinued. Two years later the New Bedford and Providence lines suspended operations. In that same year there was a sitdown strike on the railroad's Fall River Line. The strike plus mounting losses caused the New Haven to discontinue the famous old line in July 1937. The Fall River Line's four side wheelers, the *Plymouth*, the *Providence*, the *Priscilla*, and the *Commonwealth*, were sold for scrap.

After the New Haven Railroad bowed out of the Sound's steamship service, a few proud boats were still plying between New York and New England, among them the *Meteor*, formerly called the *Chester W. Chapin*, acquired by the Colonial Line in 1938. She went to New Bedford during the summer season and in winter sometimes replaced the line's other vessels, the *Arrow* and the *Comet* on the Providence run when the latter ships were withdrawn from service for repairs. The *Meteor* remained in the Sound until 1941. When the United States entered World War II, the government took her, along with the *Arrow* and a number of other former Long Island Sound boats including the *Richard Peck* of the Messick excursion line running to Playland and Bridgeport during the summer. The *Richard Peck* wound up in Newfoundland during the war, then transferred to Virginia, serving as a passenger ferry for the Pennsylvania Railroad until finally condemned in 1953. She was replaced on the Playland run by the *John Messick*, formerly called the *Naushon*, used in the New London service during the winter.

In their heyday the proud vessels which plied the waters of the Sound carried millions of passengers with an amazing safety record. To be sure, there were occasional accidents right down to the 1930s. In January 1935, for example, the Sound steamer *Lexington* of the Colonial Line, bound for New England with a large group of Brown and Harvard university students returning to school after the Christmas holidays, sank in the East River. Unlike the crew of the ill-fated *Slocum*, which had gone down there in 1904, the crew of the *Lexington* was well trained. They managed to provide for the safety of all but a handful of people. The crew's job was probably facilitated by the fact that no panic broke out among the passengers. The latter conducted themselves well, listening to the directions of the ship's officers, their excitement calmed by the strains of music coming from the orchestra on board until almost the last minute. Despite the loss of several of the *Lexington's* passengers, regularly scheduled passenger service on the Sound came to an end with a surprising safety record, even though many of the boats ran in winter when the Sound is sometimes choked with ice.

Ferries on Long Island Sound

Another type of passenger service reached its apex in the decade before World War II, namely, the cross-Sound ferries. With as good a safety record as the Sound steamers, they were usually taken out of service during the winter months. Even the few operating most of the year suspended service in January and February. This was especially the case with the older ferryboats on the Sound such as the *Rye Cliff*, running between Rye and Seacliff around 1908. Since this ferry provided cross-Sound service at a time when automobiles were still a novelty, many of the passengers took horses and carriages aboard. Depending upon the tide, the trip took about an hour, a relatively short time when one considers that today's traveler between Rye and Seacliff can spend much longer than an hour just stuck in traffic on the Long Island Expressway. Another of the early twentieth-century ferryboats plied between College Point, Queens, and Manhattan. One of the boats of the College Point Line caught fire in the East River in 1905 at approximately the same spot where the *Slocum* sank. Fortunately, the fire was contained.

The Rye–Bayville ferry service, also known as the Rye–Oyster Bay ferry, was one of the oldest on the Sound, dating from 1739 when a ferry ran from the mouth of the Byram River in Rye Town to Matinecock Point on the North Shore of Long Island. In the 1700s ferries ran from New Rochelle to Hempstead Harbor and from New London to Sag Harbor with regular passenger service.

With auto ferry service beginning in the twentieth century, drivers liked to cross the Sound between Greenwich and Oyster Bay. Legend has it that the proposal to establish a regular crossing here originally came from Andrew Carnegie in the 1890s. He had to cross the Sound at this point in a hurry and had to pay $250 for the steamer that took him and his carriage across. But the fare for the typical traveler in 1917 was considerably less than what Mr. Carnegie paid: two dollars for car and fifteen cents for every passenger. Men in military uniform could cross free. The Greenwich–Oyster Bay ferry made six round trips a day, but this rigorous schedule was too taxing for the crew, with the result that one Sunday in July 1917 the boat departed an hour late on its 4 P.M. run to Oyster Bay. The delay would have passed unnoticed except for the fact that one of the irate passengers was none other than Theodore Roosevelt. T. R. arrived at the Greenwich ferry dock at 3.50 P.M. Waiting impatiently, he finally questioned the manager about the long delay. He received the explanation that the numerous trips exhausted the crew, causing the inconvenience. T. R. promptly suggested that one of the daily round trips be eliminated in order to lessen the strain and enable the crew to handle the crossings according to schedule. Heeding Roosevelt's recommendation, the management revised the schedule, omitting the little-patronized noontime trip.

In the early twenties one could reach the opposite shore of the Sound from

Oyster Bay, a focal point of ferry crossings, by going to Greenwich, Rye, or New Rochelle. The Westchester division in a few years merged with the Greenwich line mainly because motor traffic headed across the Sound came from the east, benefiting the ferry at Rye more than that at New Rochelle. Another line, between Clauson Point in the Bronx and College Point in Queens, served the more westerly sections of the Sound, operating on a twenty-four-hour basis. Still, the real traffic was in the east where Bridgeport and New London maintained busy crossings.

Encouraged by the popularity of the eastern ferries, the Long Island Rail Road decided to run auto steamers from Sag Harbor to Greenport on the one hand and New London on the other. In 1925 Stamford joined the list of ferry terminals by establishing a line to Huntington with a boat capable of carrying forty cars and one thousand passengers, extending it the following year by an additional crossing from Stamford to Oyster Bay. Not to be set back in the competition, the Greenwich–Oyster Bay line added another boat to its service, increasing the trips to twenty a day with a total capacity of seven hundred cars. The mid-twenties may be considered the peak period in the history of the Long Island Sound ferries. But with the coming of the Great Depression, a rapid decline set in, aggravated by the building of bridges, the mortal enemies of ferries everywhere.

First came the Triborough Bridge, opened in 1936, facilitating travel to the western end of Long Island; this was followed by the Whitestone Bridge in 1939. By allowing the motorist to bypass the island of Manhattan, the latter bridge became extremely popular. Unlike most Long Island ferries, it was ready and waiting to take the motorist from the mainland to Long Island or vice versa at any time of night or day, winter or summer. Although more will be said about bridges in chapter 11 it should be kept in mind that the forerunners of the Long Island Sound bridges, those ubiquitous and reliable ferryboats, were an important part of the Sound's history. With the exception of the New London–Orient Point and Bridgeport–Port Jefferson ferries, which are still running, Long Island Sound ferries, like the great steamships, sailed off into history in the 1930s. Yet as long as they existed, the ferries fulfilled a purpose. They not only transported people and cars across the Sound but, particularly during the depths of the depression, they enabled an autoless passenger to have an outing on the Sound.

Storms

For passengers aboard the Bridgeport–Port Jefferson ferryboat *Park City,* one particular trip across the Sound, that of September 2 1938, proved to be a memorable outing. The boat had left Bridgeport at 11 A.M. and reached Port Jefferson at 1 P.M. despite fierce winds. Assuming that it was safe to make the return trip, the

127

Park City's captain set out for Bridgeport at 2 P.M. with twenty-nine passengers, including an infant, and nine crewmen aboard. En route they encountered heavy winds and seas whipped up by a dreadful storm. Happily, the experienced captain could ride out the storm despite water in the hold and the loss of electricity. After twenty-one hours help arrived to tow the ship into Port Jefferson. In the meantime the *Park City's* sister ship the *Priscilla Alden* had left Bridgeport at 6 P.M. on the same day. Once out on the Sound, her captain, realizing the magnitude of the storm, decided to return to Bridgeport. While docking in the harbor, she suffered some damage after being struck by high waves. The ferry *Catskill*, proceeding from Orient Point to New London, rode out the storm in the sheltered water in back of the north fork of Long Island. It then continued across the Sound, only to discover that the ferry dock in New London no longer existed.

Such storms of hurricane force engulf Long Island Sound every so often. The storm of 1815 is well remembered. Called the "great September gale," it caused considerable damage in the shore areas of Connecticut around New London and on the eastern end of Long Island. The record also mentions fierce hurricane-like winds devastating the Sound in the years 1821 and 1869. These were September gales. Winter brought its own surprises. The Christmas Eve storm of 1811 is believed to have been even worse than the famous blizzard of 1888. In 1910 a prolonged cold spell played havoc with shipping. The ice closed up on an entire fleet in the Sound off New Rochelle. Fourteen years later in 1934 the temperature dropped to fourteen degrees below zero, and the bays and inlets of the Sound froze solidly. One adventurous motorist drove his automobile on the ice from Manhasset to New Rochelle. On Manhasset Bay itself the big sport was attaching sleds to cars for a motorized version of "snap the whip." To have more excitement, the thrill seekers went out for nighttime driving on Manhasset harbor. The unusually cold weather resulted in more than fun and games, causing serious problems for the lighthouse keepers, some of whom lived for days in complete isolation and ran out of food.

Looking back on the unusual weather which sometimes gripped the Sound area, one realizes that the region is periodically subjected to fierce summer and autumn storms and extremely cold winters. There have also been times when the seasonal clock malfunctioned, such as the year 1816 when frost and snow occurred in June and August. When the real winter arrived, the severe weather disappeared. December 1816 was unusually mild.

Unexpected weather patterns are bad enough when they last for several days or weeks, but when a change in the weather occurs in less than an hour, the results can be disastrous. Several such meteorological changes occurred in the 1930s. In June 1930, for example, a rain and hail storm hit the Sound, disrupting regattas. Twelve boats participating the Cold Spring Harbor Beach Club's races capsized;

128

two sank and two washed ashore, but none of the thirty-six people aboard was injured. Fog, sometimes more dangerous than a storm, was the culprit in the 1936 incident involving a group of junior naval reservists whose boat the *American Boy* ran into a ledge near Tree Island. The forty-eight boys were rescued by the Coast Guard and taken to New London. A group of people cruising in the Thimble Islands off Branford, Connecticut, enjoyed equally good fortune in July 1939. The fog became quite thick, causing the skipper of the yacht to turn back. While he attempted to do so, a sudden gust of wind hit the boat and overturned it. The disaster could have been prevented if the passengers had not climbed to the upper deck, making the craft top-heavy. Fortunately, wearing life preservers, they could stay afloat until rescued. Far less fortunate was the crew of a schooner who lost their lives when a fierce storm in May 1938 drove their ship aground on No Nation Reef off City Island.

The Great Hurricane of 1938

Seldom did the area of the Sound experience such a devastating storm during recorded history. On September 21 1938 the full fury of the winds reached the eastern suburbs of the New York metropolitan area. Here the trees, uprooted and broken, lay pointing southward like dead soldiers mowed down while in close formation. At the eastern extremity of the huge cyclone revolving counterclockwise, the fierce gales whipped up thirty-foot waves in Narragansett Bay and drove the water into the narrow harbor of Providence, hurling a huge tidal mass into the downtown area of the city. The eye of this two-hundred-mile-wide turbulence rode across Long Island and the Sound lashing gigantic waves against the mainland shoreline with appalling results. The waves would carry all manner of craft torn from their moorings and deposit them wherever they happened to run aground, while the winds would lift houses off their foundations and carry them away. Through the poor visibility of the heavy rain, they would look like huge objects flying among the smaller debris.

Strange things happened. A New Haven train was halted outside Stonington by a house and a cabin cruiser on the tracks. As the heavy rain washed out an embankment, the terrified passengers saw a three-story house above them, being carried by the storm. At New London another New Haven Railroad train found its passage obstructed by a lighthouse tender. The boat remained there for weeks after the hurricane while officials argued over granting a permit to remove it. The shore line of the New Haven Railroad could not be restored to service until the second week in October.

The interruption of rail transportation intensified the economic effects of the

129

hurricane of 1938. Few people had adequate insurance coverage, and coming as it did during the depression, the storm exacerbated an already disastrous situation. Within a year of the terrible storm, however, residents of the Sound area had rebuilt many of the structures, both residential and commercial, destroyed by the hurricane.

The World's Fair in Flushing Bay

In 1939 there seemed to be a new optimism in some of the areas devastated the previous year. One such region was Long Island, as commented upon by the editor of the *Nassau Daily Review-Star* in October 1939:[5]

So far as I know there is only one place in the United States where villages are still growing into cities, doubling their populations every few years. I know of only one place where farms are being cut up into city lots on a large scale and new communities are springing up in the fields where we used to grow corn and potatoes. . . . I am talking about Long Island. It is a paradox of the troubled twentieth century that Long Island, one of the first frontiers of America, also should prove its last.

The hurricane of 1938 notwithstanding, Long Island grew rapidly as a metropolitan area in the late 1930s. During the World's Fair in 1939, the Long Island Association used the opportunity to mount an elaborate campaign to convince visitors of the island's assets.

The location of the World's Fair on Flushing Bay proved to be very advantageous for transportation by water; the fair was easily accessible for Sound yachtsmen who wished to go by boat. If you happened to be a Westchesterite without a boat, you could still sail to the fair aboard a cruiser out of Mamaroneck. The boat provided daily passenger ferry service to the World's Fair. The trip took about an hour, except for the day in May 1939 when the boat became stranded on Execution Rock. She eventually returned to Mamaroneck with everyone aboard safe.

World War II

By the time the World's Fair ended, much of the world was at war, and once the United States became involved in the conflict, pleasant boat rides on Long Island Sound became less frequent. Instead of pleasure boats using the Sound in large numbers, naval vessels plied its blue waters, continuing a tradition began much earlier. More than a decade before Pearl Harbor, the United States military used Long Island Sound as a test area for all sorts of new equipment ranging from communications devices to submarines. In May 1930 a three-day mock Battle of Long

130

Island Sound was held. The "enemy," known as the Black Fleet, departed from the naval base at Newport, Rhode Island. Their goal was to push into the eastern end of the Sound past the forts on Fishers, Plum, and Great Gull islands. Opposing the invaders were submarines and military planes from Trumbull Field, Groton, and Mitchel Field, Long Island.

The mock Battle of Long Island Sound was indicative of the lingering fear, which can be traced as far back as the American Revolution, that the shore communities might be the object of a seaborne attack. This apprehension had surfaced a number of times, with varying results. During the Spanish–American War, for example, the fear of a seaborne attack led to the erection of Fort Tyler on a part of Gardiner's Island which later became separated from the island by erosion. Built of rock brought in by boat from the Palisades, the fort was completed two years after the war. Used by rumrunners during Prohibition, Fort Tyler fell victim to neglect. When a general went out to inspect it in the 1920s, the action of the wind and waves had piled sand inside the fort, making the place look like something out of the *Arabian Nights*. The incredulous general, unwilling to believe that this was a U.S. Army fort, asked if someone was trying to play a joke on him.

By 1941 military fortifications were no longer a joking matter, nor was apprehension about an enemy attack launched from the Sound laughable. Indeed, some people, fearing enemy submarine activity, refused to venture out to the beaches, let alone onto the Sound itself. This situation became so alarming by 1942 that the managing director of the Long Island Association took to the radio to convince listeners that the beaches were safe. He pointed out that the Long Island beaches were open to the public from early morning until 8:30 P.M. in summer and that there was absolutely no truth to the rumor that people were prohibited from sailing and fishing unless they had identification cards. The Long Island Association denied just as emphatically another story that dead sailors had been found on the island's beaches. The association also denounced rumormongers who claimed that oil from torpedoed tankers covered Long Island beaches, killing the waterfowl. Twenty-five years later oil tankers did not need to be torpedoed to spread oil slicks, but that is getting ahead of our story.

No matter how safe the beaches, most Long Islanders had no time to enjoy the great outdoors. They were too busy making a contribution to the war effort. Many island residents found employment in the defense industries. By 1942 unemployment in Nassau County reached the lowest level in a dozen years. Long Islanders who had been out of work in the 1930s were happy to put in the long hours demanded by their new jobs. After all, a person could do little in his leisure time. The beaches, though open, could not be easily reached, because the manufacture of passenger cars stopped completely in 1942, and gasoline rationing began the same year. A few of the Gold Coast millionaires could still splurge in the early forties, though more

131

and more the big parties they used to give turned into benefits for the military or the Red Cross. Instant millionaires emerged on both sides of the Sound, thanks to the black market in sugar, meat, and tires. Some people shunned the black market and resorted to substitutes for various household products. The Long Island State Park Commission encouraged homeowners to use their fireplaces for heating instead of consuming precious heating oil. The commission ordered that driftwood be piled up on the beaches for people to collect for their fireplaces. Considering the gasoline shortage, one wonders how many people actually went to the beaches to gather wood. And what about those who lived in the new homes called "defense houses" built on Long Island during the war, costing between $2,750 and $4,250? How many of these dwellings had fireplaces?

To turn to a far more serious matter, what about the people who, according to a Long Island Association report, fled because of the possibility of an enemy attack? Long Island had been vulnerable at other times in history. The Revolution had produced a huge band of refugees. The fear that World War II would also result in an enemy take-over became more intense as the war dragged on. The blackouts imposed in the coastal area every evening pointed up the seriousness of this fear. Apprehension mounted on Long Island in particular because of the presumed vulnerability of any area having large defense installations. Members of various police forces on the island were removed from their normal posts and placed in defense plants as guards. The Army had to guard the Long Island Lighting Company's generating plant at Glenwood Landing, following a serious power failure in August 1942. The blackout, interpreted by some as the work of saboteurs although an investigation could produce no evidence of sabotage, interfered with work at seven defense plants.

On the other side of the Sound, attitudes and actions resembled those on Long Island. Initially workers were delighted to find jobs after the lean years of the depression. Even before the United States became involved in the war, Connecticut benefited greatly from the economic boom generated by the conflict in Europe. In 1940 the number of industrial plants in the state reached the level of 1929. The State Development Commission busily advocated that employers move to Connecticut, by pointing to the absence of a state income tax, sales tax, and dividend tax. Just as in World War I, the factories of Bridgeport and New Haven made a valuable contribution to the defense effort. Industrialized towns like New Rochelle and Port Chester along the Sound coast of Westchester did their part for America and her allies, too. Elsewhere on the Sound shore of Westchester, the citizens made other contributions to the war effort. The people of Rye for instance made contingency plans for receiving refugees from New York in the event the great metropolis had to be evacuated. Fortunately, these plans never had to be carried out, though clear and present danger did exist, if not in New York City, then certainly offshore in the

132

Sound. Boats were sunk mysteriously as late as 1945 when a German submarine appeared at the eastern end of the Sound and an American coal ship sank. The United States Navy and Coast Guard searched for the sub and after locating it near Block Island destroyed it with depth charges.

Aftermath of the War

Dramatic incidents such as this were superseded in the years after 1945 by more humdrum activities. Along with new housing developments on the North Shore of Long Island where Gold Coast mansions once stood came highways, factories, and corporate headquarters. Up and down the mainland side of the Sound, former city dwellers, whether individual families or huge corporations, found new homes. There were, of course, economic dislocations after the war, but they did not occur unexpectedly. In 1944 Governor Baldwin of Connecticut warned that new jobs would have to be found for many people and that public works programs could not be relied upon to meet this need. On Long Island the end of hostilities in August 1945 was bound to cause large-scale unemployment. Yet as late as the spring of that year, optimistic Long Islanders were thinking and talking not so much about the island's problems as about its achievements.

One writer pointed to the phenomenal growth of Nassau County, particularly the areas along the Sound. He also foresaw the comparative reduction of the size of the county by automobile transportation. Peering into the future, he concluded: "No one knows how small the county may become when the helicopter is our common means of conveyance,"[6] a fitting remark, considering that Long Island claimed to have been the cradle of the American aviation industry. In fact, the first commercial flights over the North Atlantic took off from Long Island Sound at Port Washington and landed in Ireland in 1937. Pan American maintained an airport on Manhasset Isle at Port Washington from 1937 to 1939. It was from here that Pan American clippers bound for Europe and Bermuda departed and French, German, and English planes landed. Indeed, prior to World War II Port Washington played the role of the American gateway for air traffic to Europe. Earlier in the 1930s F. Trubee Davison, the assistant secretary of the Navy, flew to Long Island from Washington in one hour and nine minutes to address the Long Island Chamber of Commerce, proposing that the United States should build more planes. Davison predicted in 1932 that in a short time it would be safer to pilot a plane through fog than a ship.

This prediction would soon come true, but the one made in the spring of 1945 about helicopters was perhaps a bit too futuristic. Instead of helicopters, autos on the shores of the Sound and boats on the waterway itself would proliferate at an

133

unbelievable rate in the postwar period and, indeed, reduce the distances around the Sound and between the Sound shore communities. As World War II ended, residents of the Sound area breathed a sigh of relief from the fear of enemy subs stealing into their sea. Little did they know that another enemy, the weekend yachtsman, was waiting in the wings to launch a massive invasion of the inland sea. Peace had arrived to be followed by a new invasion, prepared not by the chiefs of staff of an enemy nation but by the new generation of suburbanites swarming eastward, engulfing the shores of the Sound in the vast sprawl of megalopolis.

Long Island Expressway under construction near East Williston, L. I. (Lockwood, Kessler & Bartlett)

134

9

Everyman's Sea:
The Postwar Period

Mass Invasion

FOLLOWING WORLD WAR II, THE TEMPO OF THE MARCH TO THE SUBURBS PICKED up noticeably. This was the period of rapid suburbanization. Not only were individual families moving out to the Sound shore communities, but huge corporations also left the city to settle in suburbia. At the same time the masses started to invade the Sound itself with everything from power boats to Sailfish. As a result, the waterway came under heavy stress. With larger numbers of people using it for recreational purposes and living on its shores, the Sound became so polluted that by the late 1960s government officials launched a massive study to determine what could be done to prevent it from turning into a dead sea. Much more will be said about this study in chapter 10, which deals with the ecology of the Sound. Here we should perhaps raise the question of how the condition of the Sound deteriorated so fast in the postwar period.

Part of the answer, of course, is the increased population in the shore communities. This, in turn, immediately raises another question, namely, why did the towns and cities along the Sound grow so much in the past twenty-five to thirty years? In searching for an answer, one is invariably struck by the relative ease of access to the Sound area provided by the new highways and expressways built in the years after World War II. Surely we cannot blame all the Sound's problems on the four main arteries funneling the largest volume of traffic into the Sound shore communities, namely, the New England Thruway–Connecticut Turnpike, the Hutchinson River Parkway and Merritt Parkway, the Northern State Parkway, and the Long Island Expressway. On the other hand, it is conceivable that without these concrete ribbons to suburbia, at least parts of the Sound area, especially in Suffolk County and along the Connecticut coastline, might have remained relatively unattractive to suburban homeseekers.

The New England Thruway–Connecticut Turnpike

The New England Thruway had been a long time coming. Going back to the 1920s, we find Westchester County officials talking about a Pelham–Port Chester parkway to relieve traffic on the congested Boston Post Road. They even bought the property necessary for building the highway. With the coming of the Great Depression, the plan was temporarily shelved, but not forgotten. Writing about vital gaps in the New York highway system in 1940, Robert Moses said that the Pelham–Port Chester parkway was one missing link to be constructed immediately to relieve other roads, including the Hutchinson River Parkway which he characterized as "the most overcrowded artery limited to pleasure vehicles in Westchester County."[1]

In 1943 the Post-War Planning Commission called for the building of the Pelham–Port Chester "thruway." At the same time the Westchester County Board of Supervisors went to the length of offering the parkway's right of way to the state of New York for the development of the badly needed main artery. However, some questions arose in Westchester about allowing this precious property to be used for a facility permitting truck traffic. Residents of Pelham circulated a petition denouncing the prospect that construction of the thruway as planned would cause the demolition of a number of expensive homes. Similar problems arose in Connecticut where the new road, known in New York State as the New England Thruway, was called the Connecticut Turnpike. Residents of the Sound shore communities through which the road would pass voiced their displeasure at the thought of huge steel monsters on wheels disturbing the quiet of their towns as they whizzed through at sixty miles per hour. The objections notwithstanding, soon after World War II ended, a meeting was held at the offices of the Triborough Bridge and Tunnel Authority on Randall's Island to discuss the right of way of the New York State section of the proposed coastal highway.

The need for the new road must have been apparent to those attending the meeting, because both the Hutchinson River Parkway and the Boston Post Road were groaning under the strain of heavy traffic. In Westchester County alone, reported the county's park commissioner, travel in the postwar period rose 10 to 15 percent above the busiest year prior to the war. The Hutchinson River Parkway was especially crowded, and the problem became worse after the opening of the West Rock Tunnel in New Haven in 1949. This improvement in the Connecticut parkway system facilitated travel between New York and points in New England north of New Haven. It also increased the volume of passenger car traffic on the Wilbur Cross and Merritt parkways in Connecticut and the Hutchinson River Parkway in New York State. Because of the ever-growing truck traffic on the Boston Post Road, the building of a highway through the Sound shore communities became a burning problem.

Problems Created by Construction of the Thruway

Once construction started on the New England Thruway–Connecticut Turnpike, some people, including those who recognized the need for a new highway to relieve congestion on existing routes, may have had second thoughts. Included in this group were those who decried the destruction of the three-hundred-year-old Gallup Gap Bridge leading to Sherwood Island in Westport, Connecticut. Homeowners in fashionable Westport were also dismayed when they discovered possums, raccoons, and skunks, displaced by the thruway, in their backyards or swimming in their pools. In Norwalk a wildcat, also apparently a refugee from the thruway, tore up suburban lawns. And animals were not the only creatures made homeless by the thruway. People living in the downtown areas of cities through which the road passed were turned out of their homes. In Stamford, Connecticut, the big question was what to do with these people until permanent homes could be found for them. One suggestion proposed to house them in the portable makeshift quarters used for victims of the 1955 floods, but then the question became where to put the portable homes. Some city officials proposed the site of the former Quonset hut colony on Shippan Avenue. In doing so they overlooked the fashionable Shippan Point area which lay beyond the proposed relocation site.

Another problem surfacing during construction of the Connecticut Turnpike was noise. In Norwalk the clamor of pile drivers and jackhammers became so bad that citizens had to obtain a court order banning work on Sunday. Blasting also posed a serious problem. A July 1956 blast, apparently caused by negligence on the part of construction workers, injured fourteen people in Greenwich. Rocks sent flying by the explosion hit overhead power lines of the New Haven Railroad, causing extensive train delays. Two months later New Haven trains were again halted when blasting for the thruway caused a rockslide in Rye. Since the thruway ran parallel to the railroad, its construction was bound to cause some interference with the trains. But state and local officials who investigated the dynamite mishap found this type of interference completely inexcusable. There were other difficulties unavoidable under the circumstances, such as the demolition of the New Haven Railroad stations at Larchmont and Rye to make room for the thruway. Although new stations were to be built, railroad officials were upset about the delays and even more by the prospect of passengers preferring to drive to New York City on

137

the thruway rather than put up with the rapidly deteriorating service and equipment of the railroad. But some time would elapse before they would face the competition.

Construction of the Westchester County portion proved especially difficult. A deep cut had to be made through downtown New Rochelle. This caused delays, as did construction of the bridge over the Byram River in Port Chester linking the New England Thruway with the Connecticut Turnpike. On a two-day tour of his state's section of the new highway in October 1957, Governor Abraham Ribicoff of Connecticut noted that the delay on the Byram River bridge would force the Greenwich segment of the Connecticut Turnpike to remain closed, thus depriving the state of $7 million in toll revenues over the next year. On another point, the governor observed that the highway would help solve the transportation dilemma of New York State and Connecticut. He also predicted that it would have a very beneficial effect upon the economy of the Sound shore communities of Connecticut. Traveling from New Haven to Greenwich on the first day of his tour, Governor Ribicoff traversed a roadway considered the longest urban highway in the country at the time because it passed through four large cities: New Haven, Bridgeport, Norwalk, and Stamford, and a number of sizable towns. Another first for the Connecticut Turnpike, observed by the governor during his tour, was the bridge over the Quinnipiac River in New Haven, whose 387-foot main span was said to be the largest single span of continuous girder construction in the United States. On the second day of his tour, Governor Ribicoff drove east from New Haven to the Rhode Island border through an area of open space, small homes, and, with a few exceptions such as the city of New London, little industry. Ribicoff pointed out that the turnpike would bring considerable development to this region, a prediction borne out within a decade of the turnpike's opening.

The Long Island Expressway

Known officially as Interstate 495, it runs from the Queens Midtown Tunnel to Riverhead, a total distance of about eighty miles. It is one of the most crowded arterial highways in the country, not only during the rush hour and not only in one direction. It is a busy thoroughfare day and night, especially its western section that lies between the tunnel and the Queens-Nassau boundary line. On this stretch traffic funnels into it from the Brooklyn-Queens, Van Wyck, and Clearview expressways, Grand Central and Cross Island parkways, and Queens and Woodhaven boulevards.

When construction of the expressway began in 1953, the engineers envisaged a peak load of 80,000 cars a day by 1970. In 1966 the road handled 140,000 cars a day and 170,000 on some weekends. In the mid-seventies the traffic is so heavy that one stalled car will create major jams, and an accident blocking two lanes will

138

cause hour-long delays. One reason is the lack of a wide enough median strip where disabled vehicles could be pulled to free the roadway. Instead, only a narrow strip exists in the middle, and cars traveling at high speed will jump the fence if they get out of control or, if the fence is strong enough, will be hurled back into the middle of the high-speed traffic. There is no room for maneuvering. Nor is there room for putting in additional lanes, because the abutments of the bridges would not allow any widening of the roadway.

Why did the planners not build a wider expressway? In Chicago the Dan Ryan Expressway has sixteen lanes, eight in each direction, connecting the downtown Loop with the western suburbs. In the rush hour it is clogged with traffic, and the local wits call it the world's largest parking lot. Local wits on Long Island call their expressway the world's *longest* parking lot, referring to the endless bumper-to-bumper traffic jams extending for miles on its mere six lanes. The irony of the situation is that no matter how many freeways are built, the traffic always increases beyond their capacity, at least during the rush hours. And freeways capable of handling rush-hour traffic without delays would be empty most of the day—a very expensive proposition even in the wealthiest states. In this respect the problems of the expressways are very similar to the problems of the commuter railroads like the Long Island. Equipment and facilities are used only part of the day, while maintenance and services are required all the time. Moreover, in the case of new roads growth usually overtakes planning because freeways, expressways, and parkways generate new traffic with phenomenal speed often exceeding the estimates of engineers by a wide margin.

For instance, the parkways built by Robert Moses in the thirties and forties as beautifully landscaped arteries to take care of weekend pleasure traffic, became commuter lines after World War II when the suburban sprawl created a mushroom-like growth of housing developments. When the Long Island Expressway was built, reaching out to the eastern end of the island and accommodating truck traffic which the parkways do not permit, the real estate developers extended their activities into Suffolk County, stimulating not only the growth of new communities but new factory towns as well. Not only did suburbia grow by leaps and bounds, but the city itself extended farther and farther out, putting increasingly heavy loads on the traffic arteries connecting the metropolitan center with its outlying settlements. Suffolk County has now, in the mid-seventies, one million inhabitants, and this number is expected to double by 1990, perhaps even earlier. But the Long Island Expressway in its western sector where the pressure is greatest will not be able to double its capacity. There is simply no room to expand unless double-decking is resorted to. However, an elevated roadway would require an increase in the capacity of the Queens Midtown Tunnel; a third tube would have to be built under the East River, in which the traffic would alternate westward in the morning and eastward in the evening rush hour. On the other hand, the rapid industrial expansion on the island itself has brought about a new development, causing two-way rush-hour traffic mornings as well as evenings, since so many people now work in the suburbs and live in the city or on its fringes and have to go to work in the opposite direction.

The rapid industrial development of the middle and the eastern parts of the island raises another burning problem. This area of three million people, with the entire industrial establishment that supports a large part of them, finds itself more and more isolated on an island which has no connection on its eastern end with the mainland. We shall see in chapter 11 that the opposition to the Rye–Oyster Bay bridge destroyed the hope of a speedy connection with the mainland side of the Sound. Another possibility would be a bridge from Orient Point to Watch Hill, Rhode Island, connecting the several islands in between and thus forming a sort of chain across the water. A bridge built here would not only open up the New England area as a market for the island's light industries; it would also establish a by-pass route via the Verrazzano Bridge for traffic coming from New England and running in the direction of New Jersey and Pennsylvania, avoiding the Bronx and Manhattan. There are of course many obstacles in the way of such a bold plan, but growth of the metropolitan area goes on inexorably, and the problem will become more acute each year with the addition of more cars and trucks and buses to the traffic choked highways of the island.

The acute controversy over bridges, expressways, and transportation of people and goods in general has so far established a trend toward solving the problem by rapid mass transit. This would provide separate lanes for express buses on highways and modernization of the existing commuter railroads so that trains running at one hundred miles per hour or even faster could travel on them. Masses of people could in this way be carried to the metropolitan area where modern transfer stations to subway lines could distribute them quickly to their individual destinations. Dispersal of population by the creation of new self-contained cities could also alleviate the problem. But New York City as a national and world center will always attract organizations, institutions, industries, and a multitude of people as daily commuters or seasonal visitors. And New York's problem will, in one way or the other, be the problem of the area around the Sound and of the Sound itself. Growth of such a dynamic center as New York City cannot be stifled, and growth of such dimensions will always impinge on the environment, raising a myriad of ecological problems.

The life of a huge metropolis like New York City may be compared to the heart of an organism pumping the blood outward to the lungs to get rid of carbon dioxide and then, revitalized by a fresh dose of oxygen, pumping the fresh blood into all the vital organs to carry nourishment to the cells. The daily commutation of four million people to New York City might be compared to the pulsation of

140

the heart. The commuters, like the blood stream, are "pumped" out to the suburbs where they can relax and get some fresh oxygen so the next morning they may return revitalized, capable of doing their daily work at the highest rate of efficiency. This socio-biological syndrome should be a top priority and should be considered the responsibility of society itself. Commutation in particular and transportation in general should be considered public services and their cost should be borne by the whole region and not only by the commuter. Only a collective effort can solve this immense problem which affects the growth of the region and in a more general sense the growth of the nation. Long Island Sound wedged as a waterway, water resource, and recreational area into this immense dynamic growth requires special consideration in order to prevent it from going down under the heavy pressure produced on all sides by growth.

Effect of the Expressways

In the wake of new highways bringing greater numbers of people to the Sound shore communities came radical transformations along the waterway. For the smaller communities there was suburbanization on a scale undreamed of in the past. Areas east of New Haven on the mainland and east of Huntington on Long Island experienced unprecedented growth. In older, more settled Sound shore communities on the North Shore of Long Island and in Westchester and Fairfield counties on the mainland, some of the remaining large estates were divided into building lots for expensive smaller homes. Where this occurred, older residents decried the end of the good old days and, on the North Shore of Long Island where development was particularly intense, some oldtimers learned to despise the land speculators and real estate agents who saw gold in every inch of waterfront land. And in fact, with the rapid development of Long Island in the years after World War II, it was apparent by the late 1960s that the suburban dream had turned a bit sour. High-intensity home building and the concommitant construction of schools, shopping centers, etc. to serve the expanding population had transformed much of the island into a megalopolis, or spread city. The rural charm had disappeared even in Suffolk County. Although development was concentrated in Nassau County in the immediate postwar period, completion of the Long Island Expressway to Riverhead in 1972 opened up Suffolk County to suburban homeseekers. Indeed, for young families Suffolk was the only possible place to live in the 1960s and 1970s when the price of homes in Nassau County had soared beyond the capacity of their pocketbooks. The Long Island housing situation was so serious that the Nassau-Suffolk Regional Planning Board, a bicounty agency headed by Lee F. Koppelman, former Suffolk County planning commissioner, in its Comprehensive Development Plan called for construction of 128,500 apartments to help house some of the anticipated 2.3 million residents of Nassau and Suffolk counties in 1985. The planning board saw a particular need for multifamily housing in Suffolk County. To solve another problem, the Comprehensive Development Plan suggested the purchase of open spaces in eastern Suffolk County. Various waterfront areas, including some along the Sound, were also pinpointed for future public purchase, among them the estate of Henry S. Morgan, sold in 1973 to a developer. County acquisition of the Morgan estate had been opposed by residents of the village of Asharoken. Access to the property is achieved by driving over a winding two-lane road linking the Eaton's Neck peninsula to the rest of Long Island. Transforming the estate into a public park would necessarily lead to traffic problems, but building too many homes on the property could cause similar difficulties. The developer, however, gave assurances that a limited number of units would be constructed, in clusters, leaving wide areas of open space. The cluster principle was endorsed by the Nassau-Suffolk Regional Planning Board as a way of maximizing the use of the remaining vacant land and minimizing man's impact upon the environment.

The cluster concept endorsed for Long Island was employed in some of the mainland Sound shore communities, usually for condominiums and garden apartments. And whether tax-conscious homeowners liked it or not, apartments went up at a record rate. Some of course were very exclusive, such as Milton Harbor House in Rye, which provided not only garages for the residents' cars but parking spaces for their boats. In practically all the Sound shore towns of Westchester and Fairfield counties, one or two apartment buildings have been constructed since 1945. Some Sound cities such as New Rochelle and Stamford have had dozens of apartment buildings constructed within the past twenty years.

In Stamford a number of these apartment buildings have been put up as part of an urban development program. Although urban renewal led to construction of new housing projects in Glen Cove and Huntington, its impact was felt more on the mainland than on the North Shore of Long Island. In the Westchester County community of New Rochelle, a large section of the downtown area was redeveloped around a shopping mall and an industrial and corporate zone. Farther up the coast in Connecticut, Stamford, Bridgeport, and New Haven underwent similar changes. Stamford, for example, with the help of the federal and city governments and assistance from private sources including a group of banks and a church, had a huge downtown facelifting which included construction of oval apartment towers overlooking the Connecticut Turnpike. Apartments were also built as part of Norwalk's urban renewal, but redevelopment there goes back to the mid-1950s when the 1955 flood necessitated rebuilding part of the city. Bridgeport had a more extensive urban renewal project in the 1960s which resulted in new housing units, a shopping mall, and revitalization of the downtown area.

Mass Housing and Urban Renewal

Perhaps even more impressive in terms of revitalization was the transformation of New Haven in the 1960s. Under the leadership of Mayor Richard C. Lee, it became a model city with a massive urban redevelopment program financed in large part by the federal government. Housing for low and moderate income families and the elderly rose where slums once stood, and tenements were replaced by modern garden apartments and town houses. The downtown area came alive with a new shopping mall and hotel, while the skyline of the city was dramatically altered by high-rise office buildings. Along the Sound the Long Wharf area, once a thriving harborside commercial zone, was rebuilt, attracting modern manufacturing and business establishments. As Mayor Lee said when the project was launched:[2]

The Long Wharf Redevelopment Plan . . . is another major step in our comprehensive program for making New Haven a city free of slums and blight. The Long Wharf area is a disgrace to our city. This plan will enable us to remove this blight from our map.

On the other side of New Haven harbor, in West Haven, the Savin Rock urban renewal project endeavored to transform a decaying waterfront area, once the scene of a popular amusement park, into a pleasant housing development. Elsewhere along the Sound coast of Connecticut, in New London, Stratford, and Milford, similar projects were launched, making Connecticut the scene of most of the urban renewal along the Sound.

As far as apartment construction along the Sound is concerned, however, the distinction of having the largest single apartment complex belongs not to Connecticut but to the east Bronx where, on the shores of Eastchester Bay, an inlet of Long Island Sound, the massive project known as Co-op City was built in the 1960s under the Mitchell-Lama middle-income housing program. Constructed on the site of the defunct Freedomland Amusement Park, the new project consisted mainly of high-rise apartments. In the words of its critics, Co-op City is sterile and a disaster from the viewpoint of planning. Schools, shopping centers, and a transportation system were lacking in the development even after several thousand residents had moved in. The absence of mass transit facilities was such a persistent problem that a proposal to use the New Haven Railroad's freight line running along Eastchester Bay was revived in 1973. First considered when Co-op City was in the planning stage, the idea had to be abandoned because the East River railroad tunnels, utilized by trains on this division to reach Manhattan, could not handle the increased traffic.

The Plight of the Long Island Rail Road

After World War II the Long Island Rail Road, like the New Haven somewhat later, fell upon hard times. Up to 1937 it had operated in the black in spite of the problems caused by the Great Depression. The feverish industrial activity generated by the war relieved the situation temporarily. As the country changed over to peacetime conditions in 1945, the peculiar geographical situation determining the fortunes of the railroad brought about an acute crisis.

The general use of the automobile for commutation, facilitated greatly by the building of parkways and expressways, reduced the number of passengers, many of them also preferring the bus to the railroad. Even greater inroads were made into the freight business of the railroad by trucks, as the extension of the highway network offered much cheaper and faster door-to-door transportation. During the war light industries grew up on the island, producing goods of smaller size but of higher value not suitable for distribution by rail. Nor would the raw materials needed for this type of manufacturing, as for instance electronics, cosmetics, instruments, and light textiles, require the transportation of heavy loads best taken care of by the railroad.

All these conditions multiplying and growing rapidly in the postwar period combined to reduce the revenues of the Long Island Rail Road while expenditures due to inflation and repeated strikes increased. Deficits grew, amounting to $4–5 million a year. The management could not maintain the equipment in good operating condition, so that breakdowns increased and serious accidents and wrecks became merely a matter of time. In 1950 two trains collided head-on in Rockville Centre causing thirty-three deaths. This was a warning signal, but nobody acted. On Thanksgiving eve in 1951 one jampacked train carrying commuters homebound from New York City broke down on the open road before it reached the Jamaica station. Another jampacked train rammed it from the rear, telescoping its front car into the rear car of the stalled train. It required a huge crane to separate the two cars so that the crushed and mangled bodies of the victims could be taken out. The accident caused seventy-seven dead and four times as many wounded. People said that now at last something would have to be done.

The newspapers and magazines, reflecting the popular view, looking for an easy explanation or for a scapegoat, blamed the Pennsylvania Railroad, the parent company of the Long Island since the turn of the century. The story went that the Pennsylvania treated its ward as an orphan, overcharging for services rendered but underpaying for everything received from the Long Island. The difference amounted to an annual loss of $2 million. This calculation, however, failed to explain the rest of the $2–3 million annual loss, totaling a deficit of $4–5 million yearly. There must have been more to it, in other words, than the alleged parsimonious treatment by the parent company, an allegation which, incidentally, the Pennsylvania Railroad vehemently denied.

Investigations and Reorganization

Following the gruesome accident of Thanksgiving eve, Governor Thomas E. Dewey of New York appointed a commission of inquiry to look into the conditions of the Long Island Rail Road and to propose reforms. The investigations brought out many facts inherent in the situation of the railroad yet more or less beyond the control of the management, whether that of the Long Island or even that of the Pennsylvania. The island itself, only 120 miles in total length from Brooklyn to Montauk, is a dead-end street for any carrier. Montauk and Orient Point do not generate much traffic, nor do Sag Harbor and Riverhead. In other words, the Long Island Rail Road is not a class one carrier. It is rather a metropolitan commuter railroad and as such the largest in the country carrying at the time of the investigation and reorganization 15 percent of the total national passenger traffic. And notoriously, transportation of commuters is the most meager source of revenue in the railroad business; even though the Long Island Rail Road had 92 million such customers a year at the end of the fifties the revenues were inadequate to meet the expenses. On the other hand, the haulage of freight dwindled steadily. The western railroads with their thousands of miles of track can make money on long distance freight of huge tonnage filling mile-long strings of freight cars. It pays for them to invest in new equipment so as to be able to apply the best innovations of modern technology to streamline operations. The Long Island Rail Road on its short distances could not do much as a carrier of bulk freight.

As a legacy of the war, industries on Long Island specialized mainly in aircraft and electronics. The products of the former leave the factory on their own power, and as for electronics and other light industries, a full week's production in these factories can hardly fill a railroad car. Thus, the Long Island's freight business steadily diminished. In 1958 its revenue from freight was $12.8 million while from passengers $52.8 million, a very unhealthy ratio. In fact, the little profit made on freight had to be used to reduce the loss on commuter traffic. Moreover, the huge number of passengers carried by the railroad travel only during four hours of the day, that is, from 7 to 9 A.M. and from 5 to 7 P.M. The rest of the twenty hours of the day the equipment is largely idle, and so are the employees, who have to be paid full-time. Toward the end of the fifties, between 8 and 9 in the morning, trainloads upon trainloads of passengers arrived at the Pennsylvania Station and in Brooklyn. The empty trains could not be stored at these places, because of lack of space. They had to be sent out to the suburbs to be brought back again in time for the evening rush. Such is the way of operation of the largest commuter railroad in America.

Already before the war, an observer noted that in the Sunnyside railroad yards in Queens, railway cars were parked in such numbers that they could move a fair-sized city and that Jamaica, a railway junction twelve miles east of Pennsylvania Station, where 751 commuter trains passed each day, would become a nationally known city if it were moved to the west. But on Long Island it is only an insignificant spot known to the commuters as a transfer point where branch lines of the railroad spread out to distant parts of the island.

Looking back at this situation from the vantage point of the mid-seventies, we see that the plight of the Long Island Rail Road in the postwar period was not an isolated phenomenon. Most of the railroads of the entire northeast section of the United States are in bankruptcy brought about by the same factors which at first appeared peculiar to Long Island, namely, short distances where the competition of the car, the bus, and the truck reduced the business of the railroad to the point where it could not operate any longer without incurring heavy losses. The western railroads, operating over huge distances, even though they face the same competition including the cargo-carrying airplane, survived, being able to generate profitable revenues. The simple fact is that in the process which blighted the railroads of the northeast section of the country, the Long Island, due to its particular situation, happened to be the first in manifesting the symptoms of an economic crisis which later swept over the whole region.

Long Island's New Business Centers

Every attempt to increase passenger traffic on the Long Island Rail Road during the day failed because people prefer to use their own cars to drive from the suburbs into town when traffic on the expressways and parkways is light. Moreover, since the big department stores followed the population exodus to the suburbs and since so many different light industries relocated their New York plants on the island, it was not necessary for the new settlers to commute to New York at all. They could find shopping centers with ample parking space and even employment, also with ample parking space, right in the new suburbs. Indeed, the exodus from the city to suburbia assumed such proportions in the postwar period that local job opportunities could not absorb all comers. Many of the residents commuted to jobs in New York City. As a result, the number of passengers on the Long Island Rail Road kept increasing without providing profitable operations. Larger numbers merely meant more equipment and more employees to take care of the additional workload — four hours a day.

To illustrate the rapid increase of population and the spread of shopping

146

Crowded marina at Port Jefferson, L. I.

Bathers at Sea Cliff public beach

*Watching the ferry head towards Connecticut
from Port Jefferson, L. I.*

*Fishing from the dock, Northport, L. I.
(Photos by Allan M. Eddy Jr.)*

Oyster farm at Northport, L. I. (Photo by the author)

Oysters at oyster farm, Northport, L. I.
(Photo by the author)

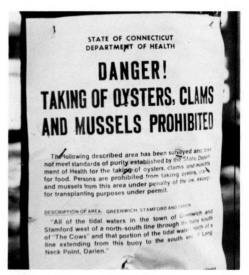

Warning sign at inlet of Sound at Cos Cob,
Connecticut (Photo by the author)

facilities and job opportunities on the island, it is sufficient to describe only two well-known examples of development, the rise of Levittown and the transformation of nearby Roosevelt Field from an airfield to a major shopping and industrial center. Both changes took place in an amazingly short time, heralding the new era that was dawning on Long Island.

The Housing Revolution: Levittown

With the outbreak of World War II, construction rapidly declined until it ceased completely. Meanwhile the population increased, so that at the end of hostilities five million new dwellings were needed to satisfy the demand for new housing. In this emergency the government acted with great foresight and resolution. Ninety-five percent of their mortgages was guaranteed to new home builders, with only five percent required as down payment, from which veterans were exempted. Repayment of the mortgages could be extended over thirty years. These generous conditions created an unprecedented building boom considered by the construction firm of Levitt and Sons, Inc., as a welcome opportunity to introduce a novel method of mass production in an industry well known for its antiquated ways.

The housing revolution was on its way. The site selected for the novel experiment was a vast stretch of land, actually twelve hundred acres of potato fields situated close to the Wantagh State Parkway. The builder aptly called the new development Levittown, a city of forty thousand people built in less than three years. Watching the operation, one could not help thinking of the assembly line at the Ford Motor Company's River Rouge plant following World War I. The only difference was that there the work came to the workman who stood at the assembly line and repeated the same operation over and over again. At Levittown the worker, or a team of workers, had to walk over to the site of a house, performing a certain operation and repeating it at the next site over and over again. Each team had its specialty. After building streets, the bulldozers and mechanical shovels prepared the foundations, and filled them with concrete brought to the site by cement mixers; then came the carpenters, who found the exact amount of precut lumber necessary for the operation at each site, left there by dump trucks. Besides the lumber there were the nails, pipes, shingles, boards, and everything necessary to complete a house. Each crew, arriving at a site, knew exactly where to look for its particular material. In this fashion the builders laid down an assembly line of prepackaged material for specially trained crewmen who, with their expertise developed to the highest degree, could complete a house every fifteen minutes. The houses sold for $7,990 and went like hot cakes.

Gloomy prophets saw a vast slum in the making, arguing that once conditions

147

in the housing industry improved, most of the people coming to Levittown would leave and be replaced by poorer people. The predictions completely missed the mark. It is true that many of the original settlers left, but the newcomers kept up, the standards, and twenty-five years later this assembly-line town has the appearance of a good middle class community. The houses became individualized by do-it-yourself improvements of their owners, and some of them, through additions, greatly increased their value even if one discounts the high prices caused by inflation.

The gloomy predictions concerning the future of housing developments produced by the assembly-line method pioneered by Levittown missed the mark because they failed to see that mass production techniques had finally invaded the housing industry. Levitt and Sons were the pioneers, and their method found imitators everywhere. Three decades later the housing industry clearly shows the trend toward mass production. and some of the largest conglomerates are moving into it with huge capital and know-how. Prefabrication and modular methods, that is, whole sections brought to the building site, are now replacing the individual single-family house builder as the supermarkets have replaced the corner grocery store. The modern way is to plan whole cities, not merely suburban developments, so that with the proliferation of new housing Long Island may become "long city," a continuous series of suburbs 118 miles long.

The Changing Scene at Roosevelt Field

The rapid increase of population brought in its wake the inevitable rise of business establishments to cater to the needs of the teeming new settlements. Roosevelt Field lies not far from Levittown, on the west side of Meadowbrook State Parkway, easily accessible from all parts of the island. The field became famous as the starting place of Colonel Lindbergh's historic crossing of the Atlantic, followed by Clarence Chamberlain and Wiley Post, and by Harold Gatty with his round-the-world record in 1931.

Because of the rapid increase of the population, the airport had to be closed. A real estate tycoon, William Zeckendorf, then hit upon the idea of developing the site into a huge shopping, office, and industrial center comprising nearly 550 acres, with parking space for 17,000 cars. Around it Nassau County, then the fastest-growing suburb of New York and at times the fastest-growing county in the whole nation, increased its population 140 percent from 1940 to 1954. The developer, looking into the immediate future of the sixties, expected a potential market of 1.6 million people within a ten-mile radius. Not only did the gloomy forecasts concerning the future of Levittown prove to be wrong, but this sanguine prediction about the growth of Roosevelt Field has been considerably exceeded by actual developments, and with the coming of expressways the ten-mile radius has doubled or tripled.

Corporate Relocation

Some of the postwar millionaires who settled on the shores of Long Island Sound built not just homes but business headquarters as well. The result of all of this was new employment opportunities for residents of the Sound shore communities, fewer of whom now had to commute to New York City. In the fashionable Westchester, Fairfield, and Nassau County towns once considered bedroom communities for New York executives, more and more people were driving to their jobs at the new corporate parks which sprang up in record numbers throughout the Sound area. Some Sound shore communities such as Greenwich, Connecticut, with its steel and glass office buildings, became mini-replicas of Manhattan. Initially, Greenwich residents welcomed the corporations with open arms. After all, what harm could a nice-looking, well-landscaped corporate headquarters do? Not much, thought Greenwich residents ten years ago. In fact, at that time homeowners were thinking in terms of the high taxes paid by the corporations. Also consoling was the thought that the new corporate neighbors employed a small number of people, emitted no noxious fumes, etc. But by the early 1970s when over one hundred corporations had relocated in Greenwich, townspeople had second thoughts. The corporate invasion threatened to profoundly alter Greenwich's image as a quiet suburban area. Thus, when the Xerox Corporation bought land in the back-country region of the town for a new headquarters, a fight shaped up over the company's request for a zoning change, enabling them to build. They argued that the property, being so close to Westchester County Airport, would be unsuitable for homes. The land, said Xerox, should be designated an airport impact zone geared to commercial rather than to residential use. Opponents of rezoning were quick to point out that an airport impact zone would open the door to a "Merritt Parkway impact zone" and a "Connecticut Turnpike impact zone" with the result that the special character of the community would be lost.

While Greenwich debated rezoning, Xerox sat it out in nearby Stamford, a Sound shore city which opened its arms to business very widely in the 1960s and early seventies. Attracting everything from research labs to cosmetic firms, Stamford became a thriving city. Farther up the coast Norwalk and Fairfield attracted some corporate giants. Norden, a division of United Aircraft, settled in Norwalk after failing to convince officials in Rye and Greenwich that it would be a desirable addition to their communities. Norden solved one of the immediate problems confronting any corporation moving to the suburbs, namely, employee housing, by helping to find quarters within easy commuting distance for its people. When

General Electric decided to build a large headquarters in Fairfield, however, the Suburban Action Institute, a private organization which seeks to abolish discrimination in suburban housing, fought the move, contending that the company's employees could not afford to live in Fairfield and would thus be unable to keep their jobs. GE's answer was that enough low-cost housing existed within commuting distance of Fairfield, so the company went ahead with its new headquarters.

Corporate relocation in the Sound shore suburbs did not always stir up the kind of opposition evident in the GE move to Fairfield. Often the trek to the suburbs was accomplished quietly. In some areas both the companies and the towns to which they moved were in agreement on the subject of relocation. Some regions even attempted to attract industry as well as "clean" corporate headquarters. This was the case on Long Island, where the Homebuilders Institute eagerly enticed companies able to provide jobs for the army of suburban home-seekers marching to Long Island in the postwar period. Land along the Long Island Expressway became the site of numerous plants and corporate headquarters, but along the North Shore glass-walled office buildings sprouted on Route 25 in formerly sleepy suburban towns. By the early 1960s only 25 percent of Long Island's work force commuted to jobs in New York City. Despite periodic recessions and cutbacks in the huge aerospace industry which employed many Long Islanders, expansion of the island's economy continued.

Westchester County also became a base for corporate refugees from New York in the years after World War II. The Sound shore community of New Rochelle underwent a renaissance in its downtown section bordering the New England Thruway. Office towers, manufacturing plants, and corporate headquarters became common sights. Although the most dramatic examples of corporate parks and individual headquarters buildings appeared away from the Sound on a strip of land bordering the Cross Westchester Expressway running from Harrison to White Plains, an area now called the "platinum mile," the new employment opportunities afforded by corporate relocation altered the commuting habits of the county to the point where large numbers of people were driving to work on highways which, with a few exceptions, had been built in the days of leisurely Sunday drives in the country.

The Crowding of Parks and Beaches

Not just her parkways came under stress but the excellent park system of Westchester was also straining under the burden of increased populaton. In the mid-fifties the parks, beaches, and pools were so overcrowded that the county's residents were complaining about not being able to get into the facilities paid for by their taxes. The problem was that the Westchester parks were open to everyone. Back

150

in 1948 the board of supervisors had considered limiting several parks, including Glen Island on the Sound, to Westchester residents, at least on weekends. The board members contended that local citizens merited special privileges, but the county parks commission questioned the legality of barring nonresidents. Then there was the difficulty of enforcing a ban. Even if ID cards were issued, there would be ways to falsify them. Besides, the state of New York might retaliate by excluding Westchester from any future programs for funding parks. For all these reasons, the ban on nonresidents was not adopted in the 1940s; but by the mid-fifties crowding had become such a serious problem that the board of supervisors again examined the residency question and decided to exclude outsiders from pools and certain beaches, including Glen Island. The county's other park on the Sound, Playland, was not included in the ban and still remains open to everyone.

Enforcement of the residency requirement is done through the issuance of permits by the Department of Parks, Recreation, and Conservation. Admittedly, the system is not perfect. Nonresidents, often from nearby Bronx County, with relatives or friends in Westchester, obtain resident cards illegally. The City of New York interpreted the residency requirement as an attempt to keep Blacks and Puerto Ricans out of the Westchester facilities. The county denied any such intention, but the borough president of the Bronx threatened to limit parking to one hour near subways in the northern part of his county, thereby making life difficult for Westchester residents who parked their cars on Bronx streets and took the subway to Manhattan. Despite the threats, the residency requirement remained, not just at county facilities but also at parks and beaches operated by the local townships. Almost without exception one has to live in a town to use the beach, all of which means that aside from Playland in Rye, access to the Sound in Westchester County is denied to everyone but residents of the municipalities which own the beaches. Inhabitants of the county not residing in a Sound shore community can use the beach at Glen Island, but this facility is hardly adequate for the multitude of Westchester residents neither belonging to private clubs on the Sound nor living in a Sound shore community.

What it all boils down to is: Who has access to our inland sea? As early as 1900 this question had posed problems for wealthy residents of Long Island's North Shore. At that time a proposal was made to purchase several hundred acres of land along the Sound to be used for a park where the people of New York City could come for a breath of fresh air. Owners of waterfront estates objected so vociferously to the mere thought of the unwashed millions invading their territory that the idea was stillborn. It was not so much a case of banning nonresidents as of not providing facilities they could use. The most exclusive areas of the North Shore simply did not have public beaches until the middle of the century, and when they were created, only residents were allowed to use them. Thus, with the exception of state parks such as Sunken Meadow and Wading River, and future parks such as Caumsett on Lloyd Neck, the metropolitan public is effectively barred from the Sound on the North Shore of Long Island.

The same holds true for Connecticut. State parks such as Sherwood Island and Hammonasset are open to out-of-state people, but practically every other strip of sand along the Sound is restricted. Some towns permit nonresidents to use their beaches for a fee often set so high as to effectively exclude outsiders. Some municipalities such as Stamford have considered allowing nonresidents to use the beaches on weekdays with the extra revenue going to acquire additional parklands. In a few cases towns, both in New York and Connecticut, have purchased private country clubs and opened them to residents. Westport, Connecticut, did this in 1960, and Rye took this route in 1965. Back in the early sixties Westchester County considered the possibility of obtaining options to purchase existing country clubs and golf courses in southern Westchester, the only large open spaces remaining. The clubs would stay for many years but would eventually be transformed into public parks. In the mid-sixties Robert Moses advocated public purchase of private clubs, especially those along the Sound, within a fifty-mile radius of New York City.

Everyman's Sea

Up to the mid-seventies nothing substantial came of these proposals, but the mere fact that public officials have begun to think along these lines demonstrates the magnitude of the problem. The increase in population of the Sound shore communities in the postwar era placed heavy demands on our inland sea. With the mass invasion of boaters and swimmers—particularly the former who pilot an unprecedented number of power boats and sailboats— the Sound certainly became everyman's sea, but at a high price, as we shall see in our next chapter. Just ask any Coast Guardsman at Eaton's Neck or New Haven. He will not only tell you about the pollution he has observed in the line of duty, but he will also recount tales of near disaster on the crowded Sound, especially on summer Sundays. According to one Coast Guardsman assigned to Eaton's Neck, the boats are so thick on weekends that you can almost walk across from Long Island to Connecticut by hopping from one boat to another.

The number of distress calls has increased sharply in recent years. Some neophyte boaters tend to overreact and summon the Coast Guard at the slightest hint of trouble. This accounts for some of the emergency calls.

Happily, the Coast Guard has been on the Sound since 1915 to assist boats in distress. Before that the Life Saving Benevolent Association of New York manned stations along the North Shore of Long Island. The first station was built at Eaton's Neck in 1850, where a lighthouse had been erected in 1799. Now the lights along the Sound are automated, and the lonely lighthouse keeper has become a memory of a bygone era. But while much of the danger of traversing the waterway is gone, problems still exist. At present more and more people residing in the affluent Sound shore communities can afford the equipment to challenge the Sound's waves and herein lies the problem. The Sound is increasingly becoming everyman's sea. True, residency requirements place much of the waterfront off limits to anyone but citizens of the Sound shore towns, but as such practices are repeatedly challenged in the courts, some changes may result. One day the Sound may truly become everyman's sea—all the more reason to clean up the waterway, thus preserving a precious natural resource. As one of the first urban seas, ringed by a densely populated area including a number of major cities, the Sound will be a prototype for the rest of the country. If it responds successfully to the heavy demands placed on it in the late twentieth century, the Sound may become a model sea. If the response is unsatisfactory, Long Island Sound may become a dead sea, a possibility unpleasant to contemplate but nevertheless deserving of serious consideration.

Sunday bathers at Sunken Meadow State Park (Photo by Allan M. Eddy Jr.)

153

Surface pollution in Port Jefferson Harbor (Photo by Allan M. Eddy Jr.)

10

The Ecology of the Sound

AFTER WORLD WAR II THE AREA OF THE SOUND UNDERWENT A RADICAL CHANGE. Before the war it was still mainly a suburban region depending largely upon the great metropolis at its western end. With the exception of the Connecticut coast up to New Haven, it was not industrialized. Westchester and the North Shore of Long Island, especially its Gold Coast section, were the domain of spacious estates, while the coastline farther east, even between the industrialized sections of Connecticut, retained its rural character. But the coming of the automobile, the building of the expressways, and the population explosion of the fifties and sixties changed this situation, so that in the seventies we see a rapid industrialization all along the shores of the Sound not only in areas close to the water's edge but farther inland as well. New population centers spring up, developing their own specialties and service industries. They are no longer tied by the umbilical cord of economic dependence to the great metropolis.

Formerly the ideal home of suburbs, this region is now assuming the aspects of rapid urbanization, functioning more as one of the geographical sections of a vast megalopolis stretching along the Atlantic seaboard from Washington to Boston. Jutting deeply into this conglomeration of suburbs, industrial sections, population centers, and rural areas, Long Island Sound is facing increasing pressures from all sides. Demands are placed upon it by an affluent population with some of the highest consumer standards in the country. Its waters are readily available and used and heavily abused for recreation, for transport, for industry, for waste disposal of all sorts.

155

Early Antipollution Measures

Under these conditions the danger of pollution was recognized very early, and the struggle to save the Sound began long before "ecology" became a household word. Already in the 1890s courses on ecology were being offered at what later became known as the Cold Spring Harbor Laboratory. In the 1920s the Long Island Chamber of Commerce undertook a tour of the island's waterways, including Long Island Sound, and pinpointed various problems affecting the condition of the Sound and the harbors on the North Shore of Long Island. The 1930s witnessed additional interest in the Sound. A New York City chemical enginering firm undertook a survey of Long Island Sound in 1930 and discovered that the bacterial content of the water was high enough to label the Sound a polluted sea. At the western end of the waterway, health officials became alarmed by these findings. The general public also became apprehensive about swimming at the popular beaches such as Glen Island in Westchester County. Bathing was actually banned at Mamaroneck by order of the Westchester County health commissioner. The problem at Mamaroneck arose from sewage dumped into the Sound before being properly treated. Similar problems would face succeeding generations of Westchesterites because, despite the construction of a multi-million-dollar modern sewage treatment facility in the 1950s, the run-off from heavy rains such as those produced by Hurricane Agnes in 1972 caused large quantities of inadequately treated sewage to be expelled into the Sound. Swimming, as a result, had to be abandoned for part of the summer of 1972, including the 4th of July weekend.

At Stepping Stone Beach in Great Neck on the other side of the Sound, the health authorities of Nassau County refused to grant permission to open the facility to the public in 1943 because of the heavy pollution of the water by raw sewage carried there from New York City by the tides. The beach, as we shall see later, has remained closed ever since. In other words, although awareness of pollution had increased between the beach closing of the 1930s and those of the early 1970s, the problem is still with us in the mid-seventies.

Congress Steps In

In view of the steadily increasing pollution threatening to turn Long Island Sound into a dead body of water, Senator Abraham Ribicoff, formerly governor of Connecticut, decided to take action. He submitted a bill to the Senate in June 1969 known as S. 2472, with the intention of establishing an intergovernmental commission to study the future of the Sound. The bill proposed that the work be undertaken by an independent commission consisting of members representing the federal,

156

state, and local governments under the sponsorship of the Senate Subcommittee on Executive Reorganization and Government Research of the Committee of Government Operations, to which the bill was referred. At the same time Congressman Lester Wolff of Nassau County introduced a similar bill, and a year later Thomas J. Meskill, then a congressman, submitted companion legislation in the House of Representatives.

Hearings were held under the chairmanship of Senator Ribicoff in New London, Connecticut, on July 7 1970, in Norwalk, Connecticut on July 8, and in Kings Point, Long Island, on July 20. The hearings were very extensive. Large numbers of experts, government officials, representatives of civic organizations, and citizens from all walks of life participated, airing grievances or pointing to problems which beset the Sound. The records, published in three volumes entitled *Preserving the Future of Long Island Sound*, form an invaluable reference. Out of this evidence one can piece together a great deal of the situation as it existed on and around the Sound in the early seventies.

The Study Plan

One of the witnesses, Frank Gregg, chairman of the New England River Basins Commission, gave the background of governmental measures that preceded the hearings. The NERBC, he explained, is a federal agency established by executive order of the president of the United States in September 1967. It functions under the authority of the Water Resources Planning Act of 1965, which set up a council of five federal agencies to coordinate programs relating to the management of water resources. The NERBC makes its reports to the president and the Congress through the Water Resources Council, which provides the basic authority.

In 1968 the state of New York decided to add parts of Long Island and Long Island Sound to the region over which the NERBC exercises authority. President Nixon confirmed this act by an executive order of April 24 1970. Meanwhile, concerned about the conditions of the Sound, the states of New York and Connecticut requested the commission to initiate prompt action to begin a study. Complying with this request, the commission, at its quarterly meeting in March 1971, set up an interagency "study plan task force" consisting of federal, state, and regional representatives who, together with an interim coordinating group of NERBC consultants, were to carry out the plan.

The study was to last three and a half years and was to be assisted by two interim advisory committees: a citizens' group to assure public participation and representatives of the scientific community of the region. It became evident at the hearings on the bills submitted to Congress by Senator Ribicoff and Congressmen Wolff and

Meskill, that the NERBC would obviate the necessity of setting up an entirely new agency as envisaged by the bill's sponsors. The NERBC was well prepared to carry out the study and draft plans for implementation by the public authorities. While the commission proposed to begin work in fiscal 1972, Senator Ribicoff and Congressman Wolff insisted that the study begin as soon as possible because the condition of the Sound was deteriorating steadily and there was no time to lose. With a supplementary appropriation of $100,000, the Water Resources Committee requested the commission to begin the study on August 1 1971. Work went ahead satisfactorily, and on that date the commission published a preliminary "Plan of Study: Long Island Sound Regional Study" for Water Resources Committee review and transmittal to the appropriations committees of the Congress. In the introduction the Plan of Study stated that the investigations of the commission would encompass[1]

the water and related land of Long Island Sound, from Throgs Neck to the Race. In Westchester County, Bronx, Queens and on Long Island, the land area of the study region will be determined by the height of land separating the Long Island Sound drainage from non–Long Island Sound; in Connecticut, the boundary is based on the jurisdictional lines of the five coastal regional planning agencies.

Further, the plan defined the purpose of the study:[2]

To recommend ways through which water and related land management can provide an environment of clean water, open space and beauty that enriches human dignity and enjoyment while maintaining solid economic opportunity for the some 11 million people within the region.

The stage was now set for a major research and planning project. Meanwhile the public hearings went on in the summer of 1970. The purpose of these hearings, as Senator Ribicoff explained, was as follows:[3]

I have always felt that when you have hearings in Washington you get technical and you get experts who come and testify but you don't have the thinking and experience of the people back home; and it would probably be to the advantage of the Congress if they would have more hearings back home giving the people an opportunity to be heard, because there is a great reservoir of intelligence in these hearings which we don't use, and it's well and good to deal with all of the big problems of international and national society; basically you can't solve them until you solve the problems back home first. . . .

The Major Pollutants

There are six major ecological threats to the existence of Long Island Sound as a national asset: oil spills and leaks, thermal pollution, dredging and dumping, destruction of coastal wetlands, discharging of raw sewage, radioactive pollution from atomic power plants. Aside from these specific factors menacing this waterway wedged in so closely between highly urbanized and industrialized areas, there are a variety of minor threats which, in their totality, constitute a serious danger equal to any of the above mentioned factors.

Oil Leaks and Spills

It may not be apparent to the sun-bathers who line the beaches along the Sound from May to October that huge quantities of petroleum move through Long Island Sound throughout the year en route to refineries and distribution centers on the North Shore of Long Island, on the Connecticut coast, and on the rivers which flow through Connecticut and empty into the Sound. Although the Sound has always been an economic artery, its use as a gigantic conveyor belt for oil has increased since the end of World War II. The influx of population in the Sound shore region and the ever-increasing use of automobiles help to explain the big rise in oil shipments. Given the wealth of the Sound shore communities and the tendency for the families who dwell in them to have more than one car, this is a real problem. Moreover, with more people living along the Sound, many of them in modern dwellings equipped with the latest appliances, there is an increased demand for electricity. The electricity itself may be considered clean energy, but to produce it the power companies must have oil to run their generators. With more people demanding more electricity to run their air conditioners, color TVs, hair driers, electric tooth brushes, etc., the problem grows. Tremendous quantities of oil must be shipped through the Sound to enable the power companies like Long Island Lighting Company (LILCO), Connecticut Light and Power, and Consolidated Edison to supply their customers. Some of the companies have developed elaborate techniques for transferring the oil to storage facilities at the power plants. LILCO, for example, has a floating terminal in the middle of the Sound where the oil is transferred to a pipeline from tankers coming up from Venezuela. Everything is handled in the middle of the Sound, including customs and immigration matters, thereby eliminating the need for the huge tankers to dock in Manhattan before proceeding to their Long Island destination.

The great danger with oil tankers, whether those supplying LILCO or anyone else, is that they are subject to serious accidents. In January 1971, for example, a tanker ran aground on a rock ledge at the mouth of New Haven harbor, spilling 385,000 gallons of light fuel oil and creating an oil slick three miles long and one and a half miles wide. The property of Humble Oil and Refining Company, which has a facility in New Haven, the ship had previously run aground off Block Island in the summer of 1970. The day after the 1971 accident, the slick had become a thin film extending ten miles along the Connecticut coast. Since it was winter, the threat to the beaches appeared less important than it might have been in summer. But residents of East Haven complained about the black ice and the stench created

159

by the oil. Conservationists were also alarmed. The Connecticut Board of Fisheries and Game reported that there were as many as five thousand ducks in the area and that 30 to 40 percent of them might die. Scientists from the Federal Environmental Protection Agency's Water Quality Office sent to study the situation found the immediate damage slight but said that the spill, like the one which had occurred at Buzzard's Bay in 1969, might eventually harm vegetation and waterfowl in the affected area.

The EPA investigated another oil spill off the Connecticut coast in March 1972, when a tanker in the vicinity of Fishers Island dumped sixty thousand gallons of oil into the Sound. The oil covered forty-five square miles, and washed ashore near Niantic, Connecticut, coating waterfowl in the area. The accident, caused when the tanker hit Bartlett's Reef, duplicated another one that had taken place in the same area in January 1969 when a barge ran aground on the reef, spilling oil over a large area of the Sound. Investigation revealed that the captain of the tanker involved in the 1972 accident had turned over the command of the ship to a first mate licensed to command ships in the Sound only between the lighthouse at Execution Rock and Stratford Shoals, not on the eastern portion of the waterway where Bartlett's Reef is located. Moreover, there were too few licensed officers aboard the vessel, an offense for which the owner of a tanker can be fined five hundred dollars.

The March 1972 spill alarmed conservationists and fishermen in Connecticut so much that the state used containment booms to close the Niantic River in order to prevent the oil from harming the river's shellfish beds. Governor Meskill of Connecticut raised the possibility of state licensing of tankers, a suggestion which produced an outcry from the oil industry. The governor also demanded that the U. S. Department of Transportation place the sea lanes farther out in the Sound.

The big Connecticut oil spill of March 1972 occurred at a time when Long Islanders were considering the possibility of dredging Port Jefferson harbor in order to permit the entry of larger tankers to supply LILCO and the Consolidated Petroleum Company. One argument against the dredging advanced by the United States Army Corps of Engineers, which has the final say in matters affecting navigable waters in this country, was that a major accident had occurred in the harbor in recent months. In January 1972 an oil tanker, only nine months out of the shipyard, broke in half. For some inexplicable reason the accident happened *after* its cargo had been pumped out of its huge hold. Anyone can imagine the magnitude of the catastrophe if the split had occurred *before* unloading—which only shows the types of dangers threatening the Sound.

Dredging and Dumping

Although the Army Corps of Engineers manifested concern over the Port Jefferson harbor question, there were times when the Corps of Engineers seemed either in-

160

sensitive or oblivious to the fate of Long Island Sound. For some fifteen years the corps renewed at three-month intervals a permit allowing Pfizer, Incorporated, a large drug company in Groton, Connecticut, to dump wastes into Long Island Sound. Pfizer claimed that the wastes produced in the manufacture of antibiotics were nontoxic. The case came to light in April 1971 during an Environmental Protection Agency conference in New Haven, Connecticut. One purpose of the conference was to set timetables for stopping contamination of the Sound. The conferees recommended the establishment of a joint water quality program for the entire Sound. But they recognized that large companies like Pfizer could get around strict antidumping laws by funneling their wastes into municipal sewer systems. Since the refuse act of 1899, which is the standard piece of legislation against polluters, does not apply to liquid discharges from municipal sewers, little could be done to safeguard the Sound against a determined polluter. As things turned out, however, Pfizer agreed to stop dumping one million gallons of waste per month into the Race. The company instead applied to the Army Corps of Engineers for a permit to discharge its waste into the ocean!

The Corps of Engineers again found itself on the hot seat in July 1972 over its plan to dredge New Haven harbor and dump the sludge out in the Sound. Also questioned was the United Illuminating Company's intention to remove sludge from the site of its new power plant in New Haven and deposit it in the Sound. People involved in the shellfish industry in the New Haven area were very alarmed by the dumping proposals. They claimed that the oyster business would be ruined by deposition of large quantities of sludge in the water. According to a study issued by the National Marine Fisheries shortly before the start of the New Haven dredging controversy, the dumping of sludge into the New York Bight, the waters around New York Harbor, had had a very detrimental effect upon marine life. The number of species of fish had declined, and many of the lobsters and crabs were diseased. This report in combination with the active opposition of the Sierra Club was enough to obtain a court order against the dredging at least until the Corps of Engineers could file an environmental impact statement explaining the effect which the dredging and dumping would have upon the Sound.

An environmental impact statement does not prevent the intended action but it does delay it for a while, thereby permitting the environmentalists to marshal additional forces. Sometimes the environmentalists can be very successful. In 1959 they won a suit to stop the dredging of Cold Spring Harbor. Two years earlier Norwalk residents made known their opposition to the dredging of a new boat channel. Particularly opposed to the project proposed by the Manhattan Sand and Gravel Company were the oystermen, who claimed that dredging would have a very detrimental effect upon their livelihood. In 1970 when the U. S. Dredging Corporation sought to remove soil from the bottom of the Sound between Ram and Chimons islands in the Norwalk Island chain, the Association to Save the Norwalk Islands fought the project, claiming that a channel between the islands was unnecessary. Senator Abraham Ribicoff of Connecticut supported the Association's stand, contending that[4]

you don't have to be a marine or civil engineer to know that when that much earth is taken from between the two islands, erosion will follow. Nature's course will be altered and maybe we will have seen the last of Ram and Chimon islands.

Destruction of Wetlands

So far we have been dealing with direct pollutants which add toxic ingredients to the waters and endanger marine life. An indirect threat to the ecosystem of the Sound and the propagation of life in it is the menace of land-fills. They eliminate shallow waters on the shore and estuaries or marshlands, areas commonly referred to as wetlands. Their importance may not be so obvious as that of clean waters in the Sound. Yet much marine life originates in these shallow waters, and wildlife thrives in them. If all marshes were filled in, wildlife would totally disappear from the Sound, and marine life in it would be badly curtailed.

Gently flowing currents in these shallow waters, carrying microscopic nourishment, were the typical breeding grounds for oysters, clams, scallops, and mussels, all precious seafood for which, as we have seen, the Sound acquired world fame in the good old days when civilization had not yet reached out to these areas. Today these industries have practically vanished from the shores of the Sound. Before 1920 the shellfish industry in the Sound realized an income of $20 million a year. Since then this income has declined to $1.5 million.

Shallow waters are also the breeding ground of all sorts of fish who come there to mate, leaving millions of eggs to hatch. The tiny offspring, swimming around in schools looking like small clouds, feed on the same microscopic elements which also support the crustaceans. When the tiny fish grow larger, they feed on the half-decomposed detritus of grass washed, together with algae, into shallow waters by the tides and storms. Microorganisms speed up the process of decomposition, producing at the same time vitamin B-12, an essential ingredient of the nutriment of all sorts of marine life. This is the beginning of the food chain of the sea which continues progressively as larger fish eat the smaller ones, ending in the large specimens supporting commercial fishermen and offering exciting game for sportsmen. It has been said that 90 percent of the fish caught in the sea depend for their growth in some way on marshlands and estuaries. In the deep they could not eat much before they would be eaten themselves. This is their fate if they are washed out by heavy tides or storms. In the shallows they can hide in the thick vegetation so characteristic of marshlands.

Many of these tidal marshes are estuarian, extending around the mouths of

162

rivers that carry silt from the interior, spreading it over vast areas overgrown with grass and weeds. Here marine species and wildlife breed well because the salt water is constantly mixing with the sweet water of the river, offering a double variety of nourishment for all kinds of life. These are the most valuable areas on the shore. Filling them up means choking off the life of the Sound.

Wetlands are also important feeding and resting grounds for waterfowl migrating on the flyways from north to south and reverse. As breeding grounds, these marshlands can affect the waterfowl population thousands of miles away.

If marshlands are destroyed by land-fills, the slow but steady sedimentation of silt on the shores of the oceans will take place somewhere else, clogging channels and harbors. Marsh peat, on the other hand, like a resilient matting, will resist erosion in severe storms, protecting the shoreline as a buffer.

Back in 1956 Abraham Ribicoff, at that time governor of Connecticut, became involved in a celebrated controversy over a saltwater marsh at Sherwood Island State Park in Westport. The state of Connecticut intended to use part of the marsh as a temporary storage area for a million yards of fill dredged from Long Island Sound for the Connecticut Turnpike then under construction. Once the work was completed, the storage area would become a permanent parking lot, a classic example of the steady encroachment of "progress" upon the environment. The Westport Audubon Society decided to fight the plan, contending that numerous birds during their annual migrations used the marsh and that the dredging operation would harm fish and oysters as well. The Connecticut State Shellfish Commission ordered the storage ground to be at least five hundred feet away from the shellfish beds, but the measure failed to satisfy the conservationists.

When the controversy over the Sherwood Island marsh began, Governor Ribicoff was on vacation. His spokesman said that while the state was considering a compromise that would reduce the area originally proposed for storage purposes and would filter the residue from the dredged soil to avoid contaminating the marsh, it would not seek other sources for the turnpike fill. The turnpike being so close to the Sound in the vicinity of Sherwood Island, there was no reason to get the fill elsewhere and then truck it to the turnpike. After inspecting the site Governor Ribicoff gave his approval, contending that he was putting people before birds and that visitors using the parking lot would outnumber the birdwatchers. In other words, the whole works, dredging, storing the fill, parking lot, and all would be carried out as originally planned. Alarmed by the governor's stand, the conservationists went to court and challenged the state's right to use the fill for public projects, declaring that valuable natural resources of the people of the state of Connecticut would be depleted, dissipated, and misused; shellfish and other natural wildlife in the areas where severance and extraction were to take place would be damaged and destroyed, and the value of the public lands underlying Long Island Sound would be greatly depreciated, all to the irreparable loss of the plaintiffs and the people of the state. Then, pondering the consequences of this ecological trespass by state authority, the protectors of the Sound warned that actually the bottom of the Sound is just as important a part of the surface of the earth as are our fields and forests. This bottom belonged to the people of Connecticut. If the state were allowed to sell it at Sherwood Island, what would prevent it from being sold anywhere to anyone, at any time?

But the opposition was to no avail. Seeking dismissal of the charges brought by the conservationists, the attorney general of Connecticut stated that private individuals could bring action against the state only if they suffered "some special damage that the public at large does not share".[5] The plaintiff, he elaborated, could not prove "any damages that the public itself is not sharing".[6] After the court's unfavorable decision the Westport conservationists picketed the site of the dredging. But only nature itself, which they defended with such determination, expressed its sympathy with their cause. High winds and rough seas delayed the dredging in November 1956, but not for long. The work proceeded as planned, despite the warning of a scientist from Yale University, voiced during the hearings on the project, that the dredging might lower the Sound's bottom twelve to sixteen feet. In answer the court ruled that the ecologists had failed to prove the case of irreparable damage done to the environment.

In 1956 the public was not yet fully aware of the importance of protecting the environment and of giving powerful support to the ecologists. But almost twenty years after the court ruling in the Westport marsh case, the country is much more aware of the need to preserve precious natural resources. The Westport conservationists of 1956 were the avant-garde of the movement, and since that time things have certainly changed. Now the concurrent opinion of all types of conservationists is that wetlands should be strictly protected and raised to the status of wildlife sanctuaries. With a note of alarm they point out that encroachment upon these valuable areas is constant, and so far more than 50 percent of them have been lost permanently. For, once wetlands are filled in and permanent structures are erected on them, they can never be restored to their original condition. Pollution of all sorts can be stopped, but filled-in areas are a permanent loss.

In Connecticut much of the marshland has been destroyed, mostly in Fairfield County and then progressively less going eastward. The state still has about twelve thousand acres of tidal marshes, of which only 2 percent appears to be safe from total destruction. They are owned by the state and conservation societies, but protection of the wetlands should include the mouths of the major rivers, the Connecticut, the Housatonic, the Thames, and the Mystic. They provide the fresh water which, mixing with seawater, offers the best habitat for marine life in the marshes. Without protection, great damage can be done, as happened for instance in the Mystic River estuary where millions of eggs of the winter flounder attached to the bottom were destroyed by the sediment deposited there from the construction of the bridge on Interstate 95.

In spite of their great importance, a relentless encroachment by private interests goes on reducing the remaining wetlands of the Sound's coastline. A recent case occurred at the southern end of Little Neck Bay, known as Udall's Cove, where there is a salt marsh. Although efforts had been made since 1963 to save it, part of the cove was filled up because the new guidelines of the Corps of Engineers giving them more power to review filling and bulkheading were not retroactive.

Another marshland lost to the surrounding community was situated in Kings Point in the vicinity of the Merchant Marine Academy. In the words of Mrs. Claire Stern, Executive Director of the Long Island Environmental Council, who spoke during the 1970 hearings on the Sound: ". . . the owner of the largest section of this marsh has filled and destroyed approximately 40 acres."[7] The marsh was partly in public, partly in private, ownership and lay on the Canadian flyway. But negotiations dragged on too long to preserve the marsh in its entirety. The witness continued:[8]

The people who live on Long Island, especially those on the north shore, are looking with some hope toward a method that can be devised that will assure them that somebody is watching out for the people's interest. Yet what we are finding is that in the politics of environmental quality the power rests not at all with the people but with those who have a vested interest in furthering their own economic power or, in their own political power. I cannot say that they are wrong. I know that the wrong is in the hands of those who say on the one hand we invite public participation but know perfectly well that the public has neither the time nor the money to have the sustained interest to protect and enhance the environment they so treasure.

The main trouble seems to be that there are too many planning agencies without power. The real power still rests in the hands of the zoning authorities, with the consequence that villages with two to three thousand residents can make decisions on whether a building may or may not destroy part of a marsh whose loss may permanently affect the ecosystem of a whole region. Passage of a Tidal Wetlands Act by the New York State legislature in the spring of 1973 imposed a moratorium on exploitation of marshlands, but some of the Sound shore communities, while welcoming the state's intention to preserve the wetlands, objected to the state overriding the local communities.

Thermal Pollution

Thermal pollution is caused mostly by power plants, which need large amounts of water to cool the condensers during the generating process. Northeast Utilities of Connecticut and the United Illuminating Company of New Haven and Bridgeport formed the Convex Pool to generate, transmit, and distribute electric energy to 1,160,000 customers in Southern Connecticut. They have thirteen generating stations of which six are located on or immediately adjacent to the Sound. Because of the rapid population growth and an even more rapid increase in the need for electric energy, the output of this pool will, during the seventies, increase from 4.3 million kilowatts to about 10 million kilowatts, and much of this increase will have to be located along the Sound because of the need for cooling water, at least in the immediate future. Later new generating techniques might be developed, and great advances might be made in thermal efficiencies. In the meantime the Convex Pool is working closely with Consolidated Edison and Long Island Lighting Company to coordinate programs and to exchange information resulting from environmental studies. A special undertaking carried out by the four utilities aims at selecting areas for closer investigation to determine the impact on the environment of future generating stations located on or near the Sound. These studies are indefinite in length and they continue as long as the problem exists.

As to the actual amount of water used for cooling purposes by the existing plants, the two atomic power plants at Millstone in Waterford, Connecticut, use a million gallons per minute, and the temperature differential is twenty-two degrees, that is, the water is that much warmer when it leaves the plant than when it is pumped in. This amount of warm water being discharged into the Sound raises the temperature by four degrees at a distance of one mile to a depth of ten to fifteen feet and in an area about three hundred feet wide. But the constant movement of the tides alters the picture as cooler water moves in and helps to dissipate the heat of the discharge. Based on these calculations, the need by additional power plants of average size will be about 400,000 gallons per minute for cooling purposes. Calculations regarding the future indicate that by 1990 25 percent of all the run-off from all the rivers in the northeast region of the United States will be used to cool power plants. Under these circumstances, the capacity of the Sound to absorb hot water will have to be determined, or else the Sound will become a heat sink for electric generation.

While the power utilities do raise the temperature of the water, they do not pollute, at least not with the cooling water. Before the water enters the plant it has to be filtered, or else the debris in it would enter the condensers and clog them. This is a common procedure in all steam plants. The water is also purified with chlorine, so it is discharged back into the Sound cleaner than it was when it entered the plant. The warm water leaving the plant can also be used to raise shellfish and plants, which seem to grow faster when irrigated with warm water. There are still experiments along these lines being conducted by Northeast Utilities in a quarry near its Millstone plant. A commercial venture in shellfish culture has been carried out by LILCO at Northport with great success, so that the maturing of the oyster,

166

which ordinarily takes four years, has been reduced to 2.5 years. This is an important matter, because shellfish production has been practically wiped out in the Sound by pollution.

In view of the enormous amount of water needed for cooling purposes, the question arises whether there are other means to solve the thermal pollution problem. One way is to employ cooling towers, but at the present stage of technology they can operate only with fresh water. If salt water is used, particles of the salt are released by the air that passes through the water to cool it. This is known as the phenomenon of drift, which so far has prevented the use of air towers on salt water. The other method is the use of long pipes jutting out thousands of feet, or even some miles, on the bottom. The discharge, coming out at great depth, is quickly diluted and cooled as it rises. This method could be used immediately, but in a few years cooling towers might be developed that will be able to handle salt water. In 1967 Congress passed a tax law with specific provisions for a fast tax write-off on depreciation of equipment used to prevent pollution. The purpose of the law was to accelerate the installation of new equipment that would reduce pollution of the waters and contamination of the air.

There are many objections to the building of new power installations, whether hydroelectric or steam-driven, and the objections are justified by various arguments. For instance, the construction of oil-fired generating units is blocked on grounds of air pollution and unsightliness. Nuclear plants are rejected because of excessive thermal discharges into cooling water and because of radioactive emissions. And finally, new major transmission lines are criticized on account of unsightliness. "Strangely," stated the chairman of the board of the Consolidated Edison Company of New York, "the jet engine peaking units, of which we are installing about 1.7 million KW in New York City to supply capacity in 1970 and 1971 we had planned to supply with the delayed nuclear plants, seem to have escaped serious objections from environmentalists."[9] He summed up the dilemma caused by the problem of protecting the environment by quoting as follows from an article published in the November 1969 issue of *Fortune* magazine, "A Peakload of Trouble for the Utilities," by Jeremy Main.[10]

Americans do not seem to be willing to let the utilities continue devouring these ever increasing quantities of water, air, and land. And yet clearly they also are not willing to contemplate doing without all the electricity they want. These two wishes are incompatible. This is the dilemma faced by the utilities.

Nuclear Power Plants

In recent years passions have been aroused not only by the actual building of power plants along the Sound but by contemplated nuclear installations. When Consolidated Edison of New York, for example, purchased David's Island, formerly Fort

Slocum, for a nuclear plant, the outcry among New Rochellians was great despite the obvious tax advantage of the proposed facility. They formed an organization called Citizens League for Education about Nuclear Energy to oppose Consolidated Edison's plan to build four atomic power reactors on the former army base half a mile offshore in the Sound. On May 29 1973 the organization held a meeting to listen to a scientist, Dr. Ernest J. Sternglass, professor of radiation physics at the University of Pittsburgh medical school. Using his study of radioactivity around the Shippingport nuclear station in Pennsylvania, he claimed that leukemia rose nearly 70 percent in the ten years following the start of plant operation in 1957, and other cancers increased 30 percent as against a statewide increase of only 9 percent. Con Ed officials disagreed with these allegations, citing the Health Physics Society's conclusions after an analysis by state and federal scientific experts, who found that Sternglass's statements were not substantiated.

Likewise, on the North Shore the Long Island Lighting Company's Shoreham plant aroused a considerable amount of opposition, not the least of which came from the Lloyd Harbor Study Group formed in 1967 to oppose a Long Island Lighting Company plant in their picturesque community. When hearings were held on the Shoreham plant in 1970, more than two dozen scientists appeared in behalf of the vociferous Lloyd Harbor Study Group. The organization's aim was to make Shoreham a test case in order to halt construction of nuclear power plants until such installations could tighten safety requirements. Arguing that the proposed Shoreham plant would raise the temperature of the Sound, thereby causing thermal pollution and destroying marine life in the area, the opponents of the plant clashed verbally with spokesmen for the utility, who claimed that the facility was needed because of the expected increase in the population of Suffolk by 1985. Unless the Shoreham plant was approved, said LILCO officials, there would be no reserves to meet the ever-increasing demand for electricity. As for pollution and the danger of radiation, company spokesmen said that the plant would be entirely safe. But when requested to guarantee that the radiation discharge would be 1 percent of the limit set by the Atomic Energy Commission, the company declared that it would try but could not guarantee anything.

Despite the objections raised to the Shoreham plant, the New York State Department of Environmental Conservation approved it in 1972, whereupon the Lloyd Harbor Study Group promptly filed suit against the commissioner of environmental conservation, but to no avail because the federal government gave its go-ahead to the Shoreham plant in the spring of 1973. With the last hurdle behind it, LILCO proceeded with construction.

The old adage about one man's meat being another man's poison is an accurate assessment of the nuclear power plant situation on the Sound. What the people of Shoreham and New Rochelle choked on, residents of the Connecticut town of Waterford devoured with gusto. When Northeast Utilities began building a nuclear plant there in 1966, the community welcomed the handsome increase of revenue from the company, which helped to keep real estate taxes for the average homeowner far below what they would be in nearby New London. Citizen reaction was so favorable that fish kills in the vicinity of the plant were pretty much overlooked. In other ways, too, the environmental impact of this nuclear facility and of a second one under construction was largely ignored by the locals. Expensive dwellings rose on the waterfront next to the plant, and people were oblivious of the high-voltage transmission lines. Now some residents are having second thoughts about Northeast Utilities' plan to construct at least a third and possibly as many as seven nuclear plants at Waterford, but opposition to Northeast is tame in comparison to the furor aroused elsewhere.

The contrast between one geographical region readily accepting the construction of atomic power plants and the other violently opposing it, is indeed striking. It reflects the dichotomy of attitudes concerning the theory of "acceptable risks" offered by the Atomic Energy Commission in view of the imperative need of the nation for more and more electric energy. Others, individuals as well as powerful organizations, cry out for an immediate moratorium on all further construction of atomic plants, arguing that the safety factor is inadequate and the risks unacceptable. No Solomonian judgment is possible here, because not even the Supreme Court of the United States can render a verdict concerning the possibility of accidents which might produce the so-called Chinese syndrome. This is a semi-facetious expression which means that if the cooling apparatus of the reactor, for some unexplicable and unpredictable reason, would collapse, the atomic core would immediately begin to heat up without any restraining influence, reaching unimaginable heat levels. Such an inferno would simply burn itself through all the steel and concrete of the foundation and would keep on burrowing deeper and deeper into the earth until it came to the surface on the opposite side of the globe somewhere in China. This possibility—that is, the breakdown of the cooling system—in more serious terminology is called the class nine accident, that is, the worst among nine possible mishaps. Such fierce fires in the core of the reactor could not be extinguished, and the continuing chain reaction would produce radioactive gases and fallouts of catastrophic proportions, poisoning the atmosphere, the waters, and the entire environment for hundreds of years. Such possibilities evoke in the public mind atomic bombs, pictures of Hiroshimas and Nagasakis with the mushroom cloud hovering over the site as the harbinger of apocalyptic destruction.

The public is understandably nervous, and opposition flares up again and again, while the environmental protection groups continue a systematic campaign demanding moratoriums and stricter controls. They maintain that radioactive seepage is still too great, that it threatens the health of the people and animals living in the

169

neighborhoods of plants, and that the disposal of the residual ash as a by-product of the operation poses problems for which no adequate solutions have yet been found.

One of the strongest challenges to the AEC and the utilities is a lawsuit initiated by Ralph Nader and Friends of the Earth, an organization formed to protect the environment against atomic pollution, which asked the federal district court in Washington to shut down twenty nuclear plants. Twelve other plants will be challenged individually. Among the plants mentioned in the court action is Waterford of Northeast Utilities, accepted so peacefully by the community. The reason for this environmental action is the allegation of the plaintiff that the overwhelming scientific evidence shows that "the lives of millions of people are threatened by the operations of these plants." This lugubrious view is supported by statements indicating that the shielding around the cores of these plants is inadequate to prevent the class nine accident from occurring, while the defendants—the AEC and the utilities—insist that safety criteria are constantly being reviewed.

While these lawsuits and controversies are and will be going on, attended by government agencies, representatives of the scientific world, the utilities, and finally the general public, the building of atomic power plants will necessarily continue, because a complete shutdown or even a moratorium on further construction would involve the whole country in immense complications. The energy needs are so great that a stoppage of new plant construction would necessitate rationing or cause brownouts, especially in regions of the country where energy shortages have already taken place regularly during the hot summer months. The demand is just insatiable. On the other hand, opposition to substitute forms of energy sources—that is, against oil- or coal-fired power plants—is also strong, and breakthroughs in technology are not in sight in the immediate future that would provide adequate energy and would satisfy all the critics.

Sewage

In 1941 the Great Neck Park District built Stepping Stone Park on a piece of land purchased from the Chrysler estate. From this excellent location on a clear day you can very well see the Empire State Building and now even more clearly the massive complex of high-rise apartments forming Co-op City. But the bathing beach with its sweeping view could never be used. Since 1942 the Department of Health of Nassau County has year after year refused to grant a permit necessary for the opening of the beach.

The location of Stepping Stone Park, besides being scenic, is also very strategic. In a way, one could call this area the headwaters of the Sound, for it is here, at the western end, that the East River empties into it. It is here that the composition of the waters which make up the Sound can best be analyzed.

170

Irked by the constant refusal of the Nassau County authorities to permit the opening of the Stepping Stone beach, the Great Neck Park District in 1958 hired sanitary engineers to check the water quality of the Sound; they obtained the following results. The currents of the East River flow at an average speed of 1.7 miles per hour, capable of carrying a large amount of material with them. Floats dropped at Welfare Island, for instance, arrived in the vicinity of Throgs Neck, a distance of about ten miles, in six hours. In spite of the many sewage treatment plants located in the East River, the current arriving at Great Neck is heavily laden with garbage and filth, full of bacteria known to exist in the intestines of warm-blooded animals. They are called coliform bacteria. According to the classification of the Nassau County Department of Health, bathing beaches are divided into four categories based on coliform count: excellent, good, fair, and bad. Classification of the water at Stepping Stone was always bad, exceedingly so—that is, over 240 counts per milliliter. "Year after year after year," said the chairman of the board of commissioners of the Great Neck Park District, "our count and that of the Nassau County Health Department has averaged a daily density of well over 1,000,"[11] a finding that was later confirmed by other tests. More than that, during many days of 1970 the count moved between 2,400 and 9,300, and on one day it shot up to 24,000, on another to 15,000. In other words, the maximum was a hundred times the upper limit of "bad," the last category. And this condition was not confined to the waters around the beach of Stepping Stone Park. Many shoreline homes and villages complained of similar conditions. Finally in the words of the chairman of the board of commissioners of the Great Neck Park District:[12]

There is a horrible point that I may add to these few statistics that I gave. There is a general coliform count, and part of that general count is the fecal matter, human excretion count. Sometimes the entire amount has a count, which is taken 6 to 9 inches below the surface of the water, that is composed almost exclusively of human excretion, human discharge, and you see it floating out here in the water. This is a horrible thing.

Some of the most ambitious studies of pollution in the Sound were launched by the Marine Sciences Research Center of the State University of New York at Stony Brook, Long Island. In 1971 a thirty-five-member team of scientists led by Dr. H. Grant Gross, assistant director of the center, set sail in a converted navy tugboat to carry out extensive tests. The team also studied areas where dredged materials and sewage were dumped. They concluded that the Sound is becoming a marine wasteland and recommended a ban on sewage disposal because harmful deposits had begun to build obstacles to navigation in the harbors. Among six harbors on the North Shore, they found Manhasset Bay to be the most polluted, a condition caused by New York City's discharge of sewage into the western part of the Sound. This statement confirmed the findings made at Stepping Stone Park. All told, between 1964 and 1969 two million tons of solid waste were dumped into the Sound.

A team from Woods Hole Laboratory in Massachusetts found plastic contaminating the waters of the Sound, while out on Great Gull Island pollutants were responsible for genetic defects in the terns. When the results of this study were published in *Natural History* magazine in 1971, the president of the New York State Association called for the appointment of a bistate commission to study these problems. Robert Cushman Murphy, the naturalist, wrote toward the end of 1971 that the Sound was on its way to becoming a sewer. Among the culprits responsible for this condition Murphy cited SUNY at Stony Brook! Pumping its sewage first into a small primary treatment plant, the university then dumped the effluent, still in a semiraw state, into Port Jefferson harbor.

There may be other offenders besides the university at Stony Brook and New York City. As a group, however, boat owners belong to a special category. Protected from the storms of the open sea, the Sound is an ideal place for boating. Yachts come from as far as Florida, and the number of boats of all descriptions using the "American Mediterranean" is increasing year by year. Whether large, medium, or small, they create their own sewage and their own disposal problem. A law regulating sewage disposal from boats was passed in 1967, to be implemented the following year. Various types of opposition caused a delay of two years. As the law finally became effective on May 1 1970, it was found that the pumping stations to handle the sewage did not exist, and the marina owners refused to comply with the law. The marinas are public, private, or semipublic and are not licensed. According to the law, the pumping stations, thirty six of them to be set up in the marinas, were to remove the sewage from the boats after they arrived at the docks. Their cost at that time was assumed to be one thousand dollars per station according to the figure of the New York State Health Department, and the boat owners were to finance their construction and maintenance by an annual fee.

Further delay arose because the plants already functioning dumped the sewage into the water in a semiraw condition — seemingly a general practice all around — so that the boat owners remarked, "We might just as well go out to sea," meaning evidently that they could dump it into the Sound themselves.

The Overall View

The danger to the Sound does not consist only of the effect of these six major pollutants. There is an overall aspect to the problem that should be seriously considered. Aside from the rapid increase of the population and their high standard of living which requires so much electrical energy, the fuel and recreational facilities which in turn stimulate the building of power plants, factories, and service establishments, there is also the problem of attitude, awareness, and sensitivity of the general public, without which the best measures will fail to accomplish their purpose. The problem of ecology so widely publicized and discussed now, cannot be dealt with from the point of view of its component parts alone unless the detailed discussions result in a holistic view, in other words, unless the general public receives an integrated picture of the problem in all its magnitude.

Let us take for instance the need for additional drinking water. Unless desalination makes substantial progress in the near future, much of the water of the rivers flowing into the Sound will be diverted to supply drinking water to the rapidly growing population, so that less and less sweet water will pour into the Sound. Industrialization, expanding in the wake of population growth, will also claim part of this valuable resource. As another aspect of the overall problem, scuba divers report that at places the bottom of the Sound is covered with beer cans, automobile tires, plastic containers and all sorts of garbage dumped into the water by thoughtless people who argue that one individual thing they happen to discard will certainly not pollute the immense expanse and volume of the Sound's waters. It may be true that many of these individual items do not poison the environment, especially since they disappear and remain out of sight on the bottom. But the totality of these minor infractions against the "ecological conscience" may eventually destroy our "Mediterranean," the value of which is so great exactly because it lies so close to the largest population center of the nation. For this reason, the Sound should be considered as an integral ecological system. If any part of it is damaged, the entire system suffers. Protective legislation should keep the integral nature of the problem in view, resulting in a comprehensive study. No complacent acceptance of individual violations, small in their relative importance, must be allowed or tolerated, and a firmer hand must be applied to shape the future of the Sound. For it is the cumulative effect of all the minor infractions, the chemical waste of factories, scrap, detergents, fly ash, waste paper, insecticides, etc., which, in addition to the major pollutants mentioned above, threaten the survival of the whole.

But first an inventory has to be taken, argued the witnesses at the 1970 hearings, to see where the most important and immediate danger lies and how the sources of pollution could be reduced and eventually completely eliminated. In other words, what should the authority, the obligations, and duties of the future agency be that will take care of these needs?

There are people, on the other hand, who scoff at ecological measures, urging instead the alleviation of the ills of our inner cities, the aiding of underdeveloped children in them, problems which in their view are more urgent than pursuing "fads" like environmental protection. These people lose sight of the fact that, if pollution continues unabated, there will be few places left where children of those inner cities can be taken to enjoy nature and go fishing or swimming. All this is difficult to solve in a democratic society accustomed to the free use of the outdoors considered as the common property of all. We will have to introduce restraints, and we will have to give up a few things, which will be more difficult when more people have more affluence and practically everybody has the means to seek the outdoors and its pleasures. The masses must be sensitized to the paramount need to use our resources with moderation, to enjoy with restraint and self-discipline, and to protect voluntarily.

The Bronx-Whitestone Bridge (Lockwood, Kessler & Bartlett)

174

11

Bridging the Sound

AMONG THE ECOLOGICAL PROBLEMS FACING LONG ISLAND SOUND, THERE IS NONE MORE serious, say some, than the plan to bridge the waterway. Opponents of bridges cite an impressive array of ecological arguments to support their position: damage to wetlands, noxious fumes from autos and trucks, harm to waterfowl, and interference with recreational boating. History may prove, however, that none of these arguments was sufficient to stop any bridge deemed necessary to serve the expanding population of the Sound area. It may well be that one day Robert Moses' dream bridge from Rye to Oyster Bay, or another Sound crossing, will be a reality. If Mr. Moses makes it to the bridge dedication ceremonies, he will probably say "I told you so" in reference to both the need for and the inevitability of such a crossing. As the politicians gather for the opening day ceremonies, there will probably be an ex post facto demonstration staged by antibridge groups from both sides of the Sound, but by that time protest will be meaningless. This little scenario, which may never be played out, thanks to the determined efforts of the antibridge forces, is a far cry from the opening day ceremonies for both the Bronx-Whitestone and Throgs Neck bridges.

The Bronx-Whitestone Bridge

On April 29, 1939 great celebrations were held at the official opening of the Whitestone Bridge, on both sides of the span. Days before the event Bronx and Queens county officials dashed about making last-minute preparations, and civic organizations made ready to do their part. Over one hundred such groups in the College Point and Whitestone areas of Queens announced that they would march across the span accompanied by color guards from the nearby army post of Fort Totten. Mayor Fiorello LaGuardia, one of the leading participants in the ceremonies, greeted four thousand people at the Bronx end of the bridge. Gazing up at the towers of the new structure, LaGuardia referred to them as symbols of efficiency in government and the democratic way of life. With great pride he pointed out that the $18-million cost of the bridge would be self-liquidating, placing no further burden upon the taxpayers, and noting also that the money so far had been spent wisely and honestly. No sticky-fingered politicians had slipped their hands into the bridge coffers, he assured the cheering crowd. Making another point, LaGuardia said: "Here, completed, we have one of the many monuments of an industrial country that confronted hardship by creating work. There she stands in all her beauty, awaiting dedication."[1] Robert Moses, chairman of the Triborough Bridge and Tunnel Authority, the agency that built the bridge, highlighted another of the project's assets. Never known for his modesty, he declared that the Whitestone Bridge was "architecturally the finest suspension bridge of them all, without comparison, in cleanliness and simplicity of design, in lightness and the absence of pretentious ornamentation."[2] Bronx Borough President James J. Lyons, engaging in a good-natured verbal duel with his Queens counterpart George Harvey, injected some bitter humor into the ceremony. Lyons braggingly declared: "We, in the Bronx, rejoice at the fine chance people in Queens will have to come over and see their friends and ancestors,"[3] to which Harvey promptly responded that the bridge would give Bronxites a convenient escape route from their own borough. On a more serious note, Grover A. Whalen, president of the World's Fair, which opened the same day as the Whitestone Bridge, referred to its future significance or, as he put it, "the importance of the bridge to the World of Tomorrow,"[4] one of the themes of the fair. Even as Whalen spoke, fifty thousand fair employees were rushing to complete last-minute jobs in preparation for the opening ceremonies on the afternoon of April 29 1939.

The Whitestone Bridge had been rushed to completion sixty days ahead of schedule in order to have it open in time for the fair. Many of the speakers at the bridge ceremony noted that the new span would enable Westchester and Connecticut fairgoers to by-pass crowded Manhattan en route to the Flushing Meadows fairgrounds. More importantly, the bridge formed part of an arterial highway system ringing the New York metropolitan area. Together with the Belt Parkway on Long Island and the extension of the Hutchinson River Parkway on the mainland, the Whitestone Bridge offered the upstate- or Long Island–bound motorist a convenient alternative to Manhattan traffic. It also helped to put Long Island into the mainstream of development, a fact undoubtedly recognized by the citizens of College Point and Whitestone, who decked out their communities in bunting to celebrate the opening of the bridge. Festivities in Queens began with a motorcade from Francis Lewis Park across the bridge and lasted into the night.

The Throgs Neck Bridge

In comparison with the opening of the Whitestone Bridge, the festivities marking the dedication of the Throgs Neck Bridge on January 11 1961 were underplayed, to say the least. Robert Moses, long-time chairman of the Triborough Bridge and Tunnel Authority, the agency that built the bridge, prohibited speeches. Politicians and guests simply gathered on the Bronx side of the span for the ribbon cutting, the traditional act marking the opening of a grand project. From there they proceeded across the bridge to a reception at Flushing Meadow Park, site of the 1939 World's Fair and 1964–65 fair, the latter still in the planning stage at the time of the dedication of the Throgs Neck Bridge. Just like the Whitestone, the $92-million structure symbolized a gateway to the new fair, being at the same time an important component of the arterial highway system of the New York metropolitan area. Connecting with the New England Thruway on the mainland side and the Clearview Expressway on Long Island, the bridge would become a well-traveled link channeling traffic away from the congested streets of New York City.

Yet on opening day when the traffic was so light that a lone bicyclist had the bridge to himself, there must have been those who wondered about the need for a new bridge. Back in 1955 the Triborough Bridge and Tunnel Authority and the Port of New York Authority had published a study of arterial facilities for the New York–New Jersey metropolitan area, concluding that another bridge linking Long Island with the mainland was needed. By 1960 some 33.5 million vehicles used the Whitestone Bridge annually, whereas in 1954 only 26 million had crossed it, indicating that the Throgs Neck Bridge would soon be needed.

Controversies About Other Bridges

There is a curious thing about bridges. No matter how badly they are needed, some-one somewhere will always pop up to oppose the project. In the early 1930s, for example, when Charles V. Bossert, a Brooklyn lumberman, came up with the idea of forming a private corporation to build a bridge from the Bronx to Whitestone, Queens, he provoked an immediate outcry. Although the regional plan for New York, prepared in 1928, included such a bridge in its Ultimate Rapid Transit Plan, the thought of a completely private bridge was unacceptable to many. To begin with, a special charter would have to be obtained from the New York State legislature, as well as the approval of the New York City Board of Estimate and of the War Department, the sanction of the latter being required for all structures spanning navigable waters. All told, the project would cost $40 million, including the sum for purchasing the necessary property on both sides of the bridge. The money would come from a syndicate of interested persons and from the sale of bonds. Presumably, the interested persons would include businessmen in the Bronx and Queens. In fact, Mr. Bossert, the originator of the project, declared that the bridge would have the beneficial effects of opening up unused and underutilized regions of those boroughs and providing jobs for five to six thousand men.

The Port of New York Authority, not seeing eye-to-eye with Bossert, filed a protest report. The New York State legislature, considering the bridge absolutely essential to the World's Fair planned for 1939, rejected the concept of a private bridge to Whitestone in favor of a span to be built by the Triborough Bridge and Tunnel Authority. The decision, taken in 1936, delighted Robert Moses, chairman of the authority. Business interests on both sides of the proposed span were as enthusiastic as Mr. Moses. Chamber of commerce and realty board officials spoke in glowing terms of the economic benefits the bridge would bring. Initially there was some skepticism about the need for another span across the East River, and one so close to the Triborough Bridge completed in 1936. But as Robert Moses explained:[5]
After the Triborough Bridge had more than come up to our expectations, we decided the time had come to discuss this project with the federal government. We haven't the slightest doubt of the success of this new bridge. It may be even more successful than the Triborough.

This statement, made in 1937, sounds like one of the most modest declarations ever to come from the lips of Robert Moses—perhaps too modest, considering the tremendous number of vehicles using the Whitestone Bridge today. As Long Island entered the final stage of its transformation from farmland to suburbia following World War II, the Whitestone Bridge provided access to mainland jobs for island residents. Traffic on the span increased accordingly to the point where relief was urgently needed. The Triborough Bridge and Tunnel Authority realized that something must be done and in 1955 set the wheels in motion for the Throgs Neck Bridge. As one of the first steps was to conduct an origin and destination survey, the Triborough Authority ordered postage-paid postcards to be handed out to motorists at the Whitestone and Triborough Bridge toll booths. The postcards contained the following questions: "Where did you start this trip? Where are you going? How many times have you made this trip in this direction in the last year?"[6] Motorists' responses to these queries would help the Triborough Authority plan the location of highways leading up to the new bridge. The bridge itself would be part of the federal government's interstate highway program. It was also, in the words of Robert Moses, "the first overt act in the establishment of a great outer belt arterial system."[7]

Moses' remark came at the ground-breaking ceremony for the Throgs Neck Bridge in October 1957. At the North Hills Country Club in Queens,* guests gath-

---

*Subsequently moved to Nassau County.

ered for luncheon to celebrate the occasion and via closed-circuit TV viewed the ceremony taking place on both sides of the span. Each guest received a brochure explaining Mr. Moses' approach to bridge building and denouncing the media for
favoring traffic relief in principle but being against it in practice. After the ground
breaking, construction proceeded rapidly. Like its predecessor, the Whitestone
Bridge, the Throgs Neck span was right on schedule. The caisson for the bridge,
manufactured in Yonkers and weighing fifteen thousand tons, arrived at the construction site in July 1958. The towing operation took eleven hours, since the caisson,
deemed too large for the Harlem River ship canal, had to be brought all the way
around Manhattan and up the East River. In the spring of 1960 the spinning of the
cables began; in September of that year they put the main suspension span into place,
and in the late fall work on the concrete roadway started. Early in 1961 everything
was ready for the grand opening. Despite initially light traffic, by the mid-1960s, the
Throgs Neck Bridge, like the other East River bridges, had reached its capacity. As
a remedy for this situation, Dr. Moses prescribed a new bridge to span the Sound
itself, not the East River, and between Nassau and Westchester counties.

The Problem of a Third Bridge

The relative ease with which the Triborough Bridge and Tunnel Authority had
constructed the Whitestone and Throgs Neck bridges may have deluded the learned
Dr. Moses, holder of an earned doctorate as well as many honorary degrees, into
thinking that the new Sound bridge could be put up with a minimum of difficulties.
But, as things turned out, the patient simply refused to swallow the medicine the
doctor prescribed. Thus, the Rye–Oyster Bay bridge became a bitter pill tossed
back and forth between irate citizen groups on one side and Dr. Moses and Governor Nelson Rockefeller on the other.

The big battle over the cross-Sound bridge began shaping up in the mid-sixties
after Moses announced his opposition to a proposed tristate bridge linking Orient
Point on the North Fork of Long Island with Connecticut and Rhode Island. The
idea for such an interstate bridge was not new. In 1938 the Commerce Committee
of the U. S. Senate discussed the feasibility of an eighteen-mile bridge from Orient
Point to Plum, Gull, and Fishers islands and then to either Groton Long Point,
Connecticut, or Watch Hill, Rhode Island. Such a span, argued its supporters in
1938, would aid the Long Island economy by providing access to New England
markets. Nothing came of the proposal until it was revived by a private group called
the Long Island Sound Tristate Bridge Committee in 1963. According to a study
understaken by this organization, a bridge from Orient Point to the Connecticut–
Rhode Island border would bring the dual benefits of traffic relief and economic

179

development. If thought of in terms of the interstate highway program, it would reduce travel time between Washington and Boston by two hours.

Alternative Proposals

Strange though it seems, the project revived (see chapter 3) pretty much what the directors of the Long Island Rail Road had in mind in the 1840s when the rail trip to Greenport and the subsequent steamer ride across the Sound to connect with the railroad to Boston were the shortest way to get to that city from New York. History indeed seemed to repeat itself, except for the fact that the citizens of Watch Hill, Rhode Island, a delightful resort area at the eastern end of the Sound, were not too thrilled by the proposal. Neither was Governor Chafee of Rhode Island, who allowed a bill establishing a tristate bridge study commission to pass but without his signature. Some New Yorkers, on the other hand, liked the idea of an impressive new bridge. Among them was Henry A. Barnes, Manhattan traffic commissioner, who viewed the bridge as an escape route from eastern Long Island which would, hopefully, keep some of the Long Island traffic out of Manhattan. Suffolk County executive H. Lee Dennison, sharing Barnes's view, felt that the bridge would end eastern Long Island's dead-end status. Dr. Lee Koppelman, Suffolk County's planning director, however, questioned whether the Long Island Expressway could handle the traffic generated by the new span.

Robert Moses added his skepticism about the tristate bridge. In June 1964, when he was sworn in for a sixth term as chairman of the Triborough Bridge and Tunnel Authority, Moses said that a tristate span would be impractical. Such a project, he said, could not be built with private funds, since its underutilization would not generate enough tolls to reward investors adequately. The only possibility was federal funding; but better still, said Moses, would be a bridge or fast ferry service between Oyster Bay and Westchester County in the vicinity of Port Chester. Within two months of the start of his sixth term, Moses commissioned a study of the Oyster Bay–Westchester bridge proposal. On the very day that Moses took the step, Bertram D. Tallamy, director of the study of the tristate bridge, declared that five to fifteen years of engineering and legislative work would be required before the span linking Long Island with Rhode Island and Connecticut could be built. This must have been music to Bob Moses' ears.

In announcing the feasibility study of the Oyster Bay–Westchester bridge to be made by the engineering firm of Madigan-Hyland, a spokesman for Mr. Moses pointed out that the landfall of the bridge on the mainland side would be in the Rye–Port Chester vicinity. He also noted that Moses had abandoned the idea of a ferry because the volume of traffic ten years hence would be too great. Moses' conclusions were borne out by the Madigan-Hyland report completed in July 1965.

180

The Plan of Robert Moses: A Bombshell

Preliminary findings of the Madigan-Hyland feasibility study team were incorporated in a report which Moses submitted to the New York State legislature in February 1965 when he sought permission for the Triborough Bridge and Tunnel Authority to build a cross-Sound bridge between Nassau and Westchester Counties. Legislative approval was necessary because the Triborough Authority's operations were limited to New York City. Warning that the bridge was inevitable, Moses maintained that a delay in approving the project could only increase the overall cost. As for the location of the bridge, Port Chester seemed to be the ideal site, at least according to the preliminary findings of the consultants. Property was less expensive than next door in exclusive Rye. There would also be, said the consultants, a smaller number of people dislocated. Building the bridge between Port Chester and Oyster Bay would, in the opinion of Madigan-Hyland, reduce the trip between Long Island and the mainland by twenty miles and at least a half-hour's driving time, advantages not obtainable by building another bridge across the East River closer to New York City. Moreover, the Port Chester–Oyster Bay span being farther from Manhattan than the existing East River crossings, it would alleviate the traffic load on the already overburdened highways of New York City, a point disputed by Manhattan Traffic Commissioner Barnes, who preferred the Orient Point–Watch Hill span, saying that the latter would divert traffic from New York City while the former would not.

Within two weeks of the initial announcement regarding his proposed cross-Sound bridge, Robert Moses, in the foreword to a brochure on the arterial highway program for the New York metropolitan area, stated that a cross-Sound bridge would be given priority. He added that when completed the bridge would be one of the engineering marvels of the century.

Intensive Opposition to the Moses Plan

Indeed, only a fortnight after the original announcement regarding the bridge, opposition on both ends of the proposed span was intense. Although we shall explore the views of the bridge opponents in depth later in this chapter, it is perhaps important to note here that when Robert Moses dropped the Sound bridge bombshell in February 1965, very few people were prepared for it, including, it is said, the Madigan-Hyland consultants, whose study was by no means complete at the time. Also caught off guard by the announcement were those agencies and groups in New York City which felt that the scope of the Triborough Authority's activities should not extend beyond the city limits. The New York City planning commissioner summed the whole thing up by saying that the profits of the Triborough Authority should be used for mass transit within the city and not for projects in Westchester

181

and Nassau counties. An editorial in the *New York Times* criticized the cross-Sound bridge idea for a similar reason, evoking a strong reply from Mr. Moses. In a letter to the editor he said:[8]

This Authority is a public agency and has statutory responsibility for vital metropolitan arterial, bridge and tunnel improvement . . . While this crossing lies outside the city lines, its major impact and effect relate to city traffic and the relief of the great East River bridges . . .

Undaunted by the opposition, including that of two members of the Triborough Authority who claimed that they had sanctioned merely an investigation of the bridge and not its construction, Moses went ahead with plans to introduce a bill in the New York State legislature giving the Triborough Authority the power to build the bridge. George V. McLaughlin, one of the dissenting members of the Triborough Authority, however, called for a separate authority to build the bridge. He also implied that the findings of Madigan-Hyland, a firm used by Mr. Moses for many years, were slanted. The consultants merely told Moses what he wanted to hear and this material, said Mr. McLaughlin, was promptly incorporated in an expensive illustrated brochure designed to produce the support needed for approval of the bridge project.

Perhaps approval would have been forthcoming if Governor Rockefeller had not hesitated, but in the spring of 1965 the governor insisted that no decision be made until the bridge project was thoroughly studied. On a related point, the governor noted that when the other facilities constructed by the Triborough Authority had been paid off by revenue from tolls, ownership of them passed to New York City. But how could the City of New York own a bridge linking Nassau and Westchester counties? Mr. Moses hoped to resolve this problem by having the New York State legislature extend the scope of the Triborough Authority's power but changed his mind because of the rising tide of opposition. Although a bill was introduced in the legislature in the spring of 1965 for this purpose, Moses requested Senator Bernard G. Gordon to withdraw it on the grounds that the bridge study had not been completed.

Moses Fights for His Plan

Could it be that Robert Moses was following the old dictum "One step backward and two steps forward" when he temporarily abandoned his bridge plan in March 1965? This is quite possible in view of the fact that Moses was back on the trail in July 1965 following publication of the long-awaited Madigan-Hyland report. The report noted the following reasons for building the bridge: relief of the existing East River bridges, reduction of both mileage and driving time for travelers between Long Island and the mainland, completion of a link in the expressway system ringing New York City, and aid to the commercial and industrial development of the region served by the proposed bridge. The report also recommended that the landfall on the mainland side be the Oakland Beach, Town Park area of Rye, not Port Chester as originally proposed. Commenting on the Madigan-Hyland report, Robert Moses stated: "From my point of view . . . the Madigan-Hyland conclusions are scientific but sensible, timely, ingenious and inescapable, and therefore should be sympathetically considered and promptly implemented."[9] Anticipating his critics, Moses went on to say:[10]

We shall of course hear the usual shrill voices of the critics decrying the rape of the open Sound and the end of primordial isolation. Every change, advance and pioneering effort runs into such derision. At the end the opponents are silent and rarely have the grace to admit their conversion. The critics build nothing.

Soon after the Madigan-Hyland report was made public, plans for the establishment of a new authority to build the Rye–Oyster Bay bridge were announced. Moses would be one of three members of the proposed agency, a situation which caused critics to note that the umbilical cord of the Triborough Authority would be firmly attached to the new agency. Circumventing another difficult problem, Moses proposed in December 1965 that the bridge pass through Playland, the Westchester County amusement park in Rye, rather than through the Oakland Beach and Rye Town Park as recommended by the Madigan-Hyland report of July 1965.

Additional Proposals

Shortly after Robert Moses advanced this suggestion, a new twist was added to the bridge controversy when a study group headed by Betram D. Tallamy, working for New York State, recommended a bridge from East Marion, Long Island, to Old Saybrook, Connecticut. The State of New York also looked into the possibility of building a bridge from Riverhead to Bridgeport and combining a rail link with a cross-Sound span, thereby making the bridge part of the Washington–Boston rail line. The concept was intriguing in view of the success, although short-lived, of the aforementioned Long Island Rail Road's cross-Sound service in the nineteenth century, but the experts did not think it would work. The "Feasibility Report: Highway-Railroad Crossings Suffolk County, New York, to Connecticut," by Bertram D. Tallamy Associates (February 1968) concluded that "present railroad traffic between Long Island and New England and that indicated by present development trends, would not financially support a toll railroad crossing of the Sound."[11]

Interesting though the rail idea was, perhaps the most unique scheme proposed for the Sound was that of Dr. Robert G. Gerard of the Lamont Geological Laboratory, who in 1966 suggested damming up the Sound by building a huge wall over which traffic could flow between eastern Long Island and Connecticut. A similar wall constructed at the western end of the Sound would enclose the waterway, transforming it into a fresh-water lake within seven and a half years after the water from the Housatonic and Connecticut rivers pushed out the salt water. The Sound then could provide drinking water for the New York metropolitan area. There would also be the side benefit of perpetual high tide, which yachtsmen would enjoy.

Governor Rockefeller Steps In

In the wake of the largely overlooked proposal to construct a bridge dam across the Sound came Governor Rockefeller's statement of March 1967 in favor of a bridge between Rye and Oyster Bay. At the same time the governor suggested the possibility of a second bridge across the Sound from Port Jefferson to Bridgeport. Making this announcement at hearings before the New York State Senate Finance Committee and the Assembly's Ways and Means Committee on his proposed transportation bond issue, Mr. Rockefeller said:[12]

I am aware of the fact that nobody wants any more bridges, roads or anything built in their area. However, Long Island, with its growth, must get an independent passing out to Westchester and Connecticut without going to New York.

Critics were quick to point out that the governor's new pro-bridge stand might have come about as the result of a compromise with Robert Moses. The quid pro quo would be Mr. Moses' support for the governor's plan to unify the transportation system of New York City through the establishment of a Metropolitan Transportation Authority encompassing the Triborough Bridge and Tunnel Authority. With this deal, the aging tsar of the Triborough Authority would get his Long Island Sound bridge.

Whether or not there is any truth to the story about the Rockefeller-Moses compromise, events now began to take a decisive turn. During its 1967 session the New York state legislature authorized the Metropolitan Transportation Authority to construct two Long Island Sound bridges, one from a point in Nassau County between Hempstead Harbor and Oyster Bay to the Rye–Port Chester vicinity in Westchester County, and the other between the area of Port Jefferson and the vicinity of Bridgeport. Hearings were scheduled to take place early in 1969. Meanwhile the city of Rye and eleven villages in the township of Oyster Bay filed suit charging that the passage the previous spring by the state legislature of Governor Rockefeller's mass transportation bond issue violated a section of the state constitution which declared that "no private or local bill . . . shall embrace more than one subject."[13] The law authorizing the bond issue actually encompassed bridges, highways, and subways. After a decision of the New York State Supreme Court favoring the plaintiffs' view, the New York State Appeals Court reversed the ruling, declaring that the Metropolitan Transportation Authority would not exceed its authority by building the bridge.

The Fight Intensifies

While the fate of the Rye–Oyster Bay span was being decided in the courts, the state of New York moved to establish a bistate commission with Connecticut to investigate the Port Jefferson–Bridgeport plan. When the Connecticut legislature failed to act, New York became that much more interested in building the Rye–Oyster Bay bridge. But you would never have known it during the 1970 gubernatorial campaign in New York. Governor Rockefeller, running for reelection, played down the bridge issue because of the mounting opposition. In fact, in April 1970 he ordered the preliminary work on the bridge halted while a full-scale transportation study was made, a decision immediately interpreted as a political move by the foes of the bridge.

On Long Island, in the meantime, opponents of the span succeeded in 1968 in having wetlands in the area of the bridge's landfall at Oyster Bay transferred to the Department of the Interior as a wildlife preserve. Lying athwart all possible approaches to the bridge, this extensive branch of the bay, now under the jurisdiction of a federal department, represented so far the most effective obstacle to the Rockefeller-Moses plan. Other measures were to follow in quick succession.

In 1971 both houses of the New York State Assembly approved a bill denying the Metropolitan Transportation Authority power to build the Rye–Oyster Bay bridge, but Governor Rockefeller vetoed the measure. An attempt to override the veto failed by seven votes. Since no gubernatorial veto had been overturned since 1872, the failure of the assembly should have come as no surprise, but there was an angry outcry, with opponents of the bridge charging that Speaker Perry Duryea had deliberately allowed legislators to leave the chamber in order to make it more difficult to overturn the governor's veto. The city of Rye filed suit, charging that Rockefeller's veto was invalid, coming as it did after the ten-day period normally allowed for acceptance or rejection of a bill, but the State Supreme Court dismissed the suit in October 1971. Two months later the firm of Creighton-Hamburg completed a "Comprehensive Transportation Study for Proposed Bridge Crossings," ordered by the state the previous year. The study concluded that the bridge should be built, among other reasons, because:[14]

If a new bridge is not built, traffic congestion on these bridges [Throgs Neck and Whitestone] and on the radial express ways will increase. The cost of goods movement will be likely to rise. The people on the island will become more steadily isolated.

In the wake of the Creighton-Hamburg report came a statement endorsing a Suffolk-County-to-Connecticut bridge from Dr. Lee F. Koppelman, executive director of the Nassau-Suffolk Regional Planning Board. According to Koppelman:[15]

If we talk of development of the Long Island area, namely industrial growth, the main problem for Long Island is elimination of the dead end pattern . . . In terms of the overall development of Long Island and the need to generate jobs, there is not the slightest doubt — I would opt for the Suffolk bridge. But the truth of the matter is, with the nature of bridges and highway transportation, probably within time we will have both bridges (Rye–Oyster Bay and Suffolk County–Connecticut).

The Nassau County Regional Planning Board had initially declined to support the Rye–Oyster Bay Bridge but gradually revised its position. In 1968 the board found the plan worthy of consideration, pointing at the same time to the problems which might be caused by the span.

As the Nassau-Suffolk Regional Planning Board was revising its views on the bridge, the state of New York had to reevaluate the financial aspects of the cross-Sound span. A report prepared by Merrill Lynch and made public early in 1972 indicated that revenue from bridge tolls would not be enough to pay for the span and that additional financial security would have to be offered investors. This established a sharp contrast with the conclusions of "Traffic, Earnings and Feasibility of the Long Island Sound Crossing between Rye and Oyster Bay" done by Madigan-Hyland for the State Department of Transportation in 1968. R. Q. Praeger, president of Madigan-Hyland, stated in a letter transmitting the report to the Department of Transportation that "the revised estimates indicate . . . that the proposed crossing . . . can be financed successfully by revenue bonds."[16]

Following the Merrill Lynch report more bad news appeared in the form of another bill passed by the New York State legislature in April 1972 to deprive the Metropolitan Transportation Authority of power to build the bridge. While a gubernatorial veto quickly disposed of the act, it could not overturn the official public disapproval of the project voiced by the Nassau and Westchester County boards of supervisors and the Nassau County Planning Committee in 1972.

Early in 1973 a federal judge postponed hearings on the approach roads to the span pending further study of the problem. The decision came after the federal government accepted a Draft Environmental Impact Statement submitted by the Metropolitan Transportation Authority and the New York State Department of Transportation, which contended that the adverse environmental impact of the bridge would be minimal and that the goal of the span's designer was to create "a graceful structure that will add interest to the Sound without being visually obtrusive."[17]

Interestingly, after the federal government accepted the report the New York State Department of Transportation requested the Federal Highway Administration to withdraw the Draft Environmental Impact Statement as the result of a federal district court decision of February 1973, "which required . . . that the preparation and distribution of such a statement await compliance with certain other procedural requirements of applicable federal laws and regulations."[18] A booklet containing information about the Draft Environmental Impact Statement distributed by the

Metropolitan Transportation Authority and the New York State Department of Transportation in December 1972 raised more than environmental issues. Utilizing a question-and-answer format, the publication answered the query "Why not build the bridge further east?"[19] by stating:[20]

Proposals have been advanced for various crossings of the Sound at sites from Suffolk County to Connecticut or Rhode Island . . . These crossings are all located outside the intensely developed portion of the New York region. They do not serve the heavy volume of traffic between the two shores of the Sound, nor would they divert anywhere near as many vehicles from the East River bridges as the Rye–Oyster Bay crossing. Therefore they cannot provide significant relief of congestion on the East River bridges or on the arteries leading to them.

The feasibility of a more easterly span was considered in a series of studies, among them "Feasibility Report: Highway-Railroad Crossings Suffolk County, New York, to Connecticut" done for the New York State Department of Transportation in 1968, confirming a 1965 report by Wilbur Smith Associates stating that a bridge in eastern Long Island would stimulate the growth of that area and would benefit interstate traffic in the long run. The Tallamy report of 1968 concluded that a bridge from Port Jefferson to Bridgeport would have the most immediate benefits for central Long Island, but the estimate of traffic and revenues in this survey was based on the assumption that the Bridgeport–Port Jefferson bridge would be built concurrently with or after the Oyster Bay–Westchester County span. Wilbur Smith Associates in their "Preliminary Report: Traffic and Revenues Proposed Suffolk County, New York, to Connecticut Crossing" of 1968 evaluated a number of easterly crossings and predicted that a Bridgeport–Port Jefferson bridge would bring substantially increased economic activity to Suffolk County. The "Long Island Sound–New England Bridge Study" done in 1968 for the New York State Department of Public Works by Sverdrup and Parcel, consulting engineers, pointed out that alignments for bridges between Port Jefferson and Bridgeport, Wading River and Momauguin, Riverhead and Guilford, and East Marion and Old Saybrook were structurally feasible.

To come back to the Draft Environmental Impact Statement, important to the basic questions raised by this report was the fate of the Oyster Bay wildlife sanctuary transferred to the Department of the Interior. Governor Rockefeller's plan to put bridge approach roads through this area was vetoed by the Interior Department, with the result that the state of New York had to consider alternate approaches including, it was said, one through Glen Cove. Another trump in the hands of the antibridge forces was a bill introduced in the House of Representatives by Ogden Reid of Westchester County and in the Senate by Abraham Ribicoff of Connecticut to prohibit the use of federal funds for approach roads unless both the New York and Connecticut legislatures approved the project. Then in 1973 a third bill was

187

passed in the New York State legislature to deprive the Metropolitan Transportation Authority of its power to build the bridge. To the utter amazement of both Robert Moses and antibridge forces Governor Rockefeller signed it. Although state legislators openly admitted that the favorite catch phrase in Albany, "Nothing is forever," might be applied to the demise of the bridge, the bridge issue indeed seemed to be dead by the summer of 1973. But how did it happen? Was it an attempt by Governor Rockefeller to win popular support for his reelection in 1974? Some Albany lawmakers thought so at the time, but in view of Mr. Rockefeller's resignation as governor in December 1973 perhaps we should explore other reasons, not the least of which was the determined opposition of the antibridge forces.

Escalation of Public Opposition

When the earlier spans linking Long Island with the mainland were built, citizens of the affected areas accepted the inevitable fact that the bridge was going to go through. Although construction of the Whitestone Bridge displaced seventeen families in Malba, Queens, those affected did not band together to protest the dislocation. In fact, their only objection was that they were given too short a time, ten days, to find new living quarters. After the City of New York acquired their homes for the approach to the bridge, orders to vacate were delivered to the homeowners. Most of them were caught off guard. The Malba residents did make an effort to obtain a thirty-day grace period, but that seems to have been the extent of their protest.

In the mid-fifties when the Triborough Authority was proceeding with its plans for the Throgs Neck Bridge, the protest was considerably louder than with the Whitestone span. On the Queens side of the Throgs Neck Bridge, homeowners displaced by the Clearview Expressway, an approach road to the bridge, pointed out that an alternate route along Little Neck Bay would have taken far fewer homes. Robert Moses shot back that the only route which would be approved and funded by the federal government was the one chosen for the Clearview. At a public hearing on the subject, witnesses accused Moses of being despotic and arbitrary in his choice of approach routes. Although the affected homeowners banded together in a Clearview Protest Committee, it was to no avail. A final decision on the Queens approach to the bridge was made in September 1957, and the following summer the City of New York undertook one of the biggest house movings in history, relocating 200 of the 428 homes in the path of the Clearview expressway.

Although the Clearview Protest Committee failed in its attempt to change the route of the Throgs Neck Bridge approach road, residents of Little Bay, Queens, were more successful when they complained about the racket caused by pile drivers at the bridge site. Much to the relief of area residents, the Supreme Court of Queens limited the pile driving to the hours between 7 A.M. and 6 P.M., thereby eliminating some of the noise which, according to residents, sometimes went on until 11 P.M.

The sweet smell of success savored by the people of Little Bay eluded a group of Bronx women who attempted to halt construction of an overpass for the approach road to the Throgs Neck Bridge in August 1959. When work crews arrived to begin moving a huge pile of sand which was to be used for an embankment for the overpass, a dozen women took up positions on Pennyfield Avenue. Claiming that they would not be buried alive, the women directed their protest not against construction workers but against life-sized effigies of Robert Moses, New York City Mayor Robert Wagner, and Bronx Borough President James J. Lyons.

A number of the women stated that they did not object to the bridge approach road but rather to the fact that the city had taken part of their property for the overpass without condemning their homes. The ladies felt that condemnation would have been more to their advantage since, they assumed, complete takeover of their property would have given them enough money to move away from the approach road.

Opposition in Rye

The attitude of the Bronx women would have mystified the women of Rye who were actively engaged in the fight to stop the Oyster Bay–Rye bridge in the late 1960s and early seventies. In the Bronx the goal was not to stop the bridge but to get a square deal from the City of New York and the Triborough Bridge and Tunnel Authority. If the affected Throgs Neck homeowners could have sold their property to the city, many would have gladly done so. The city was paying a good price for Throgs Neck real estate in those days—so good, in fact, that the courts looked into the matter. With cash in hand and upward mobility in mind, some of the Pennyfield Avenue families probably would have moved to the suburbs.

For most Rye residents living in the path of the Oyster Bay bridge, however, it was an entirely different story. Most of these people were very happy with their lovely suburban homes and waterfront estates. Thoughts of upward mobility rarely crossed their minds, because they had already reached the top. To some Rye residents upward mobility meant moving from suburban Westchester County with its New York State income tax to nearby Greenwich, Connecticut, with its nonexistent state income tax and comparatively low real estate taxes. But, by and large, Rye homeowners were content to stay put. Like their counterparts in Oyster Bay, they were dismayed by the thought of their property being taken for a cross-Sound bridge. Seeking to preserve the status quo, people from Rye and Oyster Bay vigorously protested the bridge from the very time the idea was put forth. Even back in February 1965, when it appeared that the bridge would be built between Port Chester and Oyster Bay, Rye residents objected to the possible effects of the span's approach roads. When the Madigan-Hyland report in July 1965 pinpointed the Milton Road, Oakland Beach area of Rye as the site of the bridge landfall on the mainland, the outcry became even greater.

By that time the opposition had also crystallized in Oyster Bay, where the Civic Association and Joint Committee on Community Rights were striving to inform the public of the harmful effects of the proposed bridge. Organizations in Oyster Bay and Rye would soon join forces to oppose the bridge, which they felt would mar the beauty of their communities, but in the meantime separate protests were staged. In July 1965 250 people attended a meeting at the Rye city hall to hear public officials denounce the bridge. The following month another meeting was held, at which Mayor H. Clay Johnson addressed the city council and the Mayor's Committee for the Preservation of Rye, telling them that the bridge plan would be killed because it contained major defects, especially with regard to financing.

In September 1965 Robert Moses went to Rye to inform the city's board of education that the bridge and its approach roads would not split school districts and furthermore would add to neither the city's tax burden nor its traffic, because the traffic generated by the bridge would remain on express highways. In December Moses was back in Westchester County addressing a group of county leaders to whom he announced a new plan which would bypass Oakland Pool, Rye Town Park, and private homes. In fact, under the revised proposal the bridge would pass over the county-owned Playland Amusement Park without sacrificing any homes. This latest plan was no more acceptable to Rye residents than previous proposals. As Mayor Johnson stated:[21]

We don't want the traffic dumped in our midst. Many of the cars would get off here and ruin our neighborhood. Traffic would be generated that would normally never come to Rye. Many of the cars going from Long Island to Manhattan and the Bronx would transfer to the new route over the bridge.

Another critic of the span, Rutherford Hubbard, chairman of the Rye Citizens Advisory Committee, objected to some advice given by Robert Moses to the people of Rye, whom the latter cautioned to be more democratic and considerate of the needs of all the people who would benefit from the bridge.

By the time the Playland version of the bridge proposal was released, opposition to any bridge affecting Rye was intense. In the fall of 1965 Rye Neighbors to Ban the Bridge collected signatures on petitions seeking federal, state, and county disapproval of the span. They also toured Westchester in a motorcade to stir up countywide opposition. Toward the end of 1965 representatives of Rye and Oyster Bay groups combating the bridge held a joint meeting at which consulting engineers hired by both communities reported that only 4 percent of the traffic from the existing East River crossings would use the new bridge.

Renewed efforts to halt the span were initiated in 1967 after Governor Rockefeller officially endorsed it. Members of the Rye City Council, the Rye Planning Commission, and the Citizens Committee for the Preservation of Rye held a hastily summoned meeting attended by approximately 150 people. Edmund Grainger, who had succeeded H. Clay Johnson as mayor, declared that the bridge was financially unfeasible, that it would funnel an unacceptable volume of traffic into Westchester's already crowded highways, and that Rye would continue to oppose the plan adamantly. Addressing himself to another point, Mayor Grainger stated in a letter to the editor of the *New York Times*:[22]

What it will do is create further traffic congestion in Nassau and Westchester, destroy two fine residential communities, do irreparable damage to the value of Long Island Sound as one of the state's finest recreational areas and add more noise and air pollution to the surrounding areas.

Other Efforts to Defeat the Bridge

The Yacht Racing Association of Long Island Sound was thinking along similar lines. Its members contended that the bridge would interfere with recreational boating and particularly with racing on the Sound. The directors of the association therefore proposed that a tunnel rather than a bridge be built, a suggestion which experts considered financially unfeasible and which Robert Moses labeled "an elaborate practical joke."[23] Moses also said, "It is diverting to know that there are still such folk. The societies for antiquities, in high cabal with Madame Tussaud and other wax works, should see that they are properly exhibited."[24]

To many residents of Oyster Bay and Rye, a tunnel linking their communities was no more acceptable than a bridge, and they said so loudly and clearly to anyone who would listen, including Governor Rockefeller when he was campaigning on Long Island in the fall of 1970. The governor, however, was so busy shaking hands and trying to win support for another term in Albany that he inadvertently signed an antibridge petition drawn up by the Committee to Save Long Island Sound, a Long Island organization with headquarters in Locust Valley. This group, as well as the Rye-based Citizens for Sound Planning, was very active in the fight against the bridge, particularly in the early seventies. By participating in protest demonstrations in Albany and maintaining a steady flow of literature, the antibridge group held up the progress which many opponents felt would transform the Sound from a great sea into two little puddles.

Other detrimental effects of the project were highlighted in a 1972 release from Citizens for Sound Planning, which bore the theme: "Before it's too late."[25] The communiqué spoke about dumping "tons of concrete and steel into a living body of water to usurp and pave over acres and acres of marsh land. This would be disastrous environmental mismanagement and irreparable once done . . ."[26] The release contended that something must be done ". . . before we awake one morning to a yellow,

191

smog-laden day, the Sound a brackish, brown pool with scores of dead fish floating in its oily base . . . stinking decay and death."[27]

Evolution of Public Protests

Why were protests of this type able to hold up a massive construction project? In part, the success of the antibridge forces in Oyster Bay and Rye stemmed from the support received from citizens and officials outside the affected communities. The fact that the mayor of Stamford, Connecticut, and six towns in the southwest Connecticut regional planning area opposed the bridge because of the increased traffic it would funnel into their region certainly helped. But then, again, Connecticut people were old hands at opposing bridges. Back in 1966 when a span from East Marion, Long Island, to Old Saybrook, Connecticut, was under consideration, there was an outcry in Connecticut, where people felt that New York state residents would be the ones to benefit from the bridge. The Antibridge Committee of Saybrook and vicinity was quickly established. Its chairman pointed out that a bridge was unnecessary, since eastern Long Island had nothing Connecticut needed. Presumably Connecticut could live without Long Island potatoes! Or perhaps there was another way to get them. When a span linking Port Jefferson with Bridgeport was considered, irate Fairfield County officials suggested substituting improved rail connections between Long Island and the mainland via the existing Hell Gate railroad bridge. Even Rhode Island residents joined the fight against what some people called the vivisection of Long Island Sound when a bridge from Orient Point to the Rhode Island–Connecticut border was proposed.

The history of bridge protests had come a long way between 1939 when Queens residents displaced by the Whitestone Bridge voiced quiet opposition to a ten-day vacate order. Between that time and the early seventies, when irate citizens from Rye and Oyster Bay marched on Albany, Americans as a whole became skilled in the language and action of protest. The black civil rights movement and the anti–Vietnam War demonstrations had schooled the entire nation in the lessons of protest, with the result that dignified matrons and sophisticated yachtsmen were willing to carry picket signs if it meant stopping the Rye–Oyster Bay bridge.

Whether or not these tactics will prove to be successful in the long run remains to be seen. There are some who predict that the Sound will be bridged not once but perhaps as many as eight times before the end of the twentieth century. With six bridges over the East River incapable of handling the traffic to Long Island, they say, isn't it time to bridge the Sound itself? An affirmative answer to this question may well be in the offing, although the downward trend in automobile usage resulting from the energy shortage may have the opposite effect. But no matter what happens, there will be those who concur with the sentiment voiced by Thomas Butler, one-time unsuccessful candidate for mayor of Rye who said: "If Mr. Moses wants to cross Long Island Sound, let him walk!"[28] Bridging the Sound—if, or as some would say, when it is done—will not be accomplished without a fight. If the Rye–Oyster Bay Bridge is permanently scrapped, the struggle may be transferred to other areas and the bridge tug of war may go marching on, perhaps for the rest of the century.

Tidal wetlands, ever important to the ecology of the Sound (Photo by Allan M. Eddy Jr.)

Pleasure boating on Hempstead Harbor (Photo by Allan M. Eddy Jr.)

Forecasting the Future of Long Island Sound

ANYONE LOOKING AT LONG ISLAND SOUND IN THE 1970s WOULD READILY ADMIT that the waterway was not the beautiful unspoiled sea Adriaen Block had seen three and a half centuries before. Even without a bridge bisecting it, the Sound of the seventies is the victim of man's abuse. Yet, with the Rye–Oyster Bay bridge seemingly dead and a government commission studying the Sound, the future of the waterway may be brighter than previously anticipated, provided concrete steps are taken to solve some outstanding problems.

The Problem of Power

Although the energy shortage which surfaced publicly in the fall of 1973 may lead to more sailboats and fewer power boats on the Sound, a development viewed by some as a positive step in halting pollution on the waterway, the problem of power may loom large in the Sound's future. Since the shortage of oil and oil derivatives will continue for the foreseeable future, it is safe to assume that atomic energy will be more widely used. Fuel for atomic power plants is readily available. The main obstacle is, as described in chapter 10, the opposition of environmentalists who, in many cases, have been able to prevent the selection of sites or the building of power plants fueled by the atom. As recently as 1973 the giant utility Consolidated Edison dropped plans to build an atomic plant on David's Island, formerly Fort Slocum, in the Sound off New Rochelle. Now the energy crisis may compel the authorities as well as the organized forces of opposition to revise their attitudes and speed up the building of atomic plants. Since atomic power generation requires large amounts of cooling water, Long Island Sound will attract companies pro-

195

ducing electric power by this method. Fossil fuel plants supplied by tankers and barges will also tend to be situated on the shores of the Sound, thereby raising an all-important question: Will the large amount of heated water discharged by these plants endanger the environment?

Assuming that the present drive for energy conservation will last only a few years, projections of energy consumption can safely be based on the increase of demand in the past up to the year 1973. The statistics of consumption going back to the year 1940 show a steady growth from 5.1 percent in the first decade (1940–50) to 5.6 percent in the second (1950–60) and finally 6.4 percent in the decade ending in 1970. According to the interim report on electric power generation by the Long Island Sound Regional Study, a further increase of 6.5 percent may be expected, after which there will be a steady decrease of growth in electric energy consumption, the assumption being that around 1980 or shortly after, the peak period will be reached, followed by a growth of about 4.6 percent until 2020.

In view of such steady growth, the disposal of heated water will create considerable problems for the biotic systems of the Sound. Unfortunately, our knowledge concerning the impact of thermal pollution on water quality is not sufficient. " . . . The unknowns still far exceed the knowns in water quality requirements — even to the experts."[1] One aspect is common knowledge. Water, as its temperature rises, holds less dissolved oxygen. At the same time the rate of oxygen utilization for breathing and other biochemical processes in aquatic species increases rapidly as the temperature of the water rises. This need doubles for each 10-degree centigrade increase. In consequence we have a double effect most unfavorable for aquatic plant and animal life. As the need for dissolved oxygen rises in warm water, the amount of oxygen decreases in proportion to the increase in water temperature. This, however, does not apply to all stages of animal life. The growth of certain species will be faster in warm water. Optimum conditions vary according to the species. If the temperature rises beyond a certain point, fish will not hatch, or only with great mortality. If a power plant for some reason has to shut down suddenly, there will be great fish kills due to the sudden cooling of the water. There are also the deleterious effects of warm water on fully grown fish, who will lose their resistance to diseases if they live constantly in warm water.

Another question is how to use the heat in the cooling water before the water is returned to its source. It could be used in agriculture for irrigation, for the hastening of germination of seeds, for extending the growing season, and in hothouses. But the use of warm condenser discharge is even more profitable in aquaculture as it is practiced now in several projects notably in the production of high-quality oysters and clams and eventually lobsters. The Long Island Oyster Farm adjacent to the Long Island Lighting Company plant at Northport

196

has been successfully producing oysters and clams for several years. Plankton and algae produced as food for aquatic animals are also being considered; this undertaking, if successful, could greatly increase the crop of seafood.

The lessons of the energy crisis, which caught the nation unaware of the danger of dependence upon foreign sources, will undoubtedly spur the construction of atomic power plants on the Sound. No more lengthy delays will be tolerated, although scientists differ, and differ honestly, about the intensity and range of radiation emanating from atomic power plants. So far, although thirty-eight of them have been constructed throughout the United States, no serious damage can be detected. On the eve of the energy crisis atomic plants produced only 5 percent of the nation's energy needs. Now plans are being made to increase this ratio to 20 percent by 1980. Such a rapid acceleration will unquestionably bring many such plants to the shores of Long Island Sound.

Since it has been estimated that by the turn of the century there will be about one thousand atomic plants in operation in this country, the Atomic Energy Commission decided to increase the safeguards that will insure the functioning of atomic power plants without accidents like overheating, the melting of shields, and radiation. The Union of Concerned Citizens based in Cambridge, Massachusetts, however, together with the Ralph Nader organization, is still critical of the regulations issued on December 27 1973 which have overruled their arguments concerning the probability of a loss-of-coolant accident. The AEC contended that interim criteria published on June 29 1971 were highly conservative. At any rate, a loss-of-coolant accident, the commission estimated, would occur once in ten million reactor years, that is in one million years of power output by ten reactors. But the Nader organization still maintains that the safeguards against a breakdown in case of an earthquake are inadequate. The controversy unquestionably will continue while the utilities, anxious to satisfy the ravenous appetite of the country for more electric energy, will build reactors at an ever-increasing tempo.

The most widely used method in generating atomic power is the so-called light water plant, which is fueled with enriched uranium 235 atoms burning in the reactor, where so-called moderators slow down the chain reaction. Thus, the heat generated by this process can be regulated. Water under high pressure circulates in the reactor, taking off the heat and producing steam to drive the electric generators. The process, in other words, is not very different from that of ordinary coal, oil or gas fired plants. In the construction of these plants, however, public concern about the possibility of the chain reaction getting out of control, or of excessive radiation, has extended the interim from the cornerstone-laying to the turning on of the current to ten years. What is needed now is to reduce lead time by three or four years in order to mitigate the energy crisis as fast as possible.

197

This can be done if mass production of a few well approved models is acceptable to the government as well as to the public and the utilities. If this is accomplished, atomic energy production may account for 50 to 55 percent of the total amount of energy generated in this country by 1990.

Meanwhile another type of reactor, operating by the breeder principle, may be completed. In this method the fuel is supplied by the more common element uranium 238. The reactor operates at a much higher degree of heat than the light water reactor, producing in the process a new element, plutonium 239, which is also fissionable and can be used in atomic reactors. The net result is that while the breeder reactor produces energy, at the same time it creates new fuel. In fact, it creates more fissionable material than it uses, hence its name "breeder." Used in this way, the available amount of uranium in this country could be stretched out as atomic fuel and thus America's independence from foreign sources of supply would be insured for a long time to come.

But the high heat generated in the breeder reactor requires the use of liquid metal—that is, sodium—which would play the role of the water used in the uranium 235 reactor. However, piping which would be absolutely leakproof under such severe stresses is difficult to construct. The removal of the plutonium, an extremely toxic material, would produce additional complications. But experimentation will continue, and within ten to fifteen years Long Island Sound may see the first models of this type operating on its shores. By that time, hopefully, the cooling off of the condenser discharge will also be perfected, even though atomic plants need much more cooling water than conventional electric plants.

In order to prevent wholesale thermal pollution, it will be necessary to recycle the cooling water with the help of either cooling towers or cooling ponds. According to the interim report on electric power generation issued by the Long Island Sound Regional Study in July 1973, a crescent-shaped lake constructed near the plant is the simplest solution. The hot water enters at one tip of the crescent while the cooled water leaves at the other. For purposes of a warm-water fishery, a smaller pre-cooling pond can be placed between the plant and the main cooling lake which, if it expands over a large enough area, could include a separate sanctuary for wildlife and another area for recreation.

The recycling of condenser discharge can also be done by circulating air through a falling spray of water, or the water may be spread over a lattice through which air moves to take off the heat. The cooled water falls into basins from which it is pumped back for reuse in the condenser. Losses from evaporation or spray draft must be replaced from the original source. Another method is the circulation of water through pipes or radiators as in automobiles. This is the dry method, much less efficient than the wet system, though there is no loss by

Hempstead Harbor from Memorial Park, Sea Cliff, L. I. (Photo by Allan M. Eddy Jr.)

Long Island Lighting Company facility at Northport, L. I. (Photo by Allan M. Eddy Jr.)

Long Island Lighting Company facilities at Port Jefferson (Photo by Allan M. Eddy Jr.)

Long Island Lighting Company facilities at Glenwood Landing (Photo by Allan M. Eddy Jr.)

evaporation; the cooling process is slow and consequently only smaller plants can use it.

With all the problems facing the power industry, an alternate method of power generation might simplify matters by avoiding the use of huge amounts of water. A number of utilities have recently decided to build a fuel cell to develop electricity without the cumbersome device of the steam-turbine-generator method. At first only a small model of 26,000-kilowatt capacity would be built. It would occupy less than a half acre of land and would supply the needs of a population of twenty thousand. In 1973 nine utilities signed contracts with the Pratt and Whitney Company to build a demonstration plant in East Hartford, Connecticut.

Fuel cells work by electrochemical reaction causing no pollution. The principle has long been known to science, but it is due to the efforts of the National Aeronautics and Space Administration (NASA) in applying the procedure for space missions that it can be applied now on a large scale. Basically, hydrogen and oxygen combine in the fuel cell to produce electric current. Construction and testing of this pilot cell in East Hartford will take four years. Afterwards, if it is successful, the participating utilities will build fifty-six additional plants. If this plan should materialize, Long Island Sound will be freed from the endless controversies caused by the expansion of power plants. The plants' need for water, their siting difficulties, and thermal and atmospheric pollution will all be eliminated by this latest achievement of American technology. Fuel cell plants, operating without noise, vibration, or pollution of any kind can be easily built and they will require neither a large area nor unsightly structures; the demonstration plant will be only eighteen feet high. The shores of the Sound may be freed from ugly structures belching forth smoke and noxious gases polluting the air and water around them. The pristine beauty of the countryside and of Long Island Sound will be preserved and sufficient electric energy produced to satisfy the needs of all.

The Danger of Oil Spills

Solving the electrical power problem, however, will not necessarily mean that the Sound will be free of pollution. Oil spills are among the worst pollutants of the Sound. The rapid increase of pleasure boats of all types, the grime washed down from the streets during storms and emptied into the Sound endanger marine life, soil the beaches, and interfere with recreation. The worst offenders are the many tankers and barges delivering oil needed for industrial use and for heating. Since the Sound is a well-sheltered body of water with a large urbanized area surrounding it and no harbors equipped to handle large tankers, it is traversed by medium and small craft carrying and unloading the precious cargo of oil.

There are twelve major ports in the Sound and two offshore handling platforms. They all require landing facilities and storage tanks. The harbors have to be dredged to provide the necessary draft, which means dumping of the contaminated silt and waste from the bottom. As the population increases, this traffic will increase, and if the standard of living continues to improve, so will the need for more oil increase, causing heavier traffic of delivery trucks on the highways. What is the solution?

Until the use of oil for industrial and heating purposes can be reduced by other means of power generation, oil deliveries on the Sound should be regulated. Eighty percent of the cargo moved on the waters of the Sound consists of oil and oil products. Spread out widely by the use of small and medium-sized tankers and barges, the method of delivery and distribution seems to be inadequate. Offshore delivery by large tankers using pipelines would free the harbors from the danger of congestion and pollution. In 1972 alone there were 187 oil spills on the Sound, five of them major accidents involving the loss of one thousand gallons or more. Concentration of deliveries at a few offshore facilities would simplify control and inspection; better training of the crews and personnel would prevent the frequent occurrence of accidents. However, unless standards are strictly enforced, even offshore facilities cannot avoid large-scale spillage. As we have seen in chapter 10, the Long Island Lighting Company handles the oil it needs for power generation at an offshore facility located far out in the Sound in order to reduce the incidence of grounding, collision, and other mishaps. Nevertheless, on July 3 1973 a tanker slipped her moorings in heavy rain, rupturing the connecting hose and causing the loss of five thousand gallons of oil. The accident led to the closing of more than ten miles of beach, including Sunken Meadow Park, which at this time of the year accommodates thousands of visitors on weekends. To prevent recurrence of such accidents, the company was going to increase its personnel and improve its weather monitoring.

As for the use of pipelines, a Nassau-Suffolk plan of the mid-sixties suggested a long-distance pipeline running the length of Long Island with tank terminals every ten or fifteen miles. It could have taken care of the entire oil and gasoline supply, eliminating heavy truck traffic on the Long Island Expressway and in the many ports along the North and South Shores. One large dock was envisaged at the eastern end of the island. But public opposition was too strong. Instead, Northville Industries built a much smaller project of twenty-three miles of pipeline carrying only number two heating oil from Port Jefferson to Plainville where the oil is stored in a 500,000-gallon tank. Even this pipeline saves 1.6 million truck miles a year on the LIE. When the pipeline opened on July 1 1973, the cost of heating oil was reduced from 14.95 to 14.55 cents a gallon, demonstrating the economies obtained by this method of distribution.

200

Sewage

Pollution problems may be partially solved by oil pipelines, but there are other pollutants in addition to oil. We have seen in chapter 10 that the main polluter on the western end of the Sound is New York City with its vast discharge of sewage. But the industrial centers and large suburbs on the Connecticut side contribute their share of sewage and industrial wastes, so much so that fishing, especially the gathering of shellfish, has had to be prohibited in large areas of the Sound. The Long Island coastline is not so badly polluted because of lack of industries and the absence of municipal sewer systems. Septic tanks are the rule on the North Shore, but these, on the other hand, threaten to contaminate the drinking water supply in Nassau and Suffolk Counties, which is drawn exclusively from wells.

In 1972 400 million gallons of sewage were pouring into the Sound from the seventy-two municipalities and from innumerable industrial and institutional sources. Primary and secondary treatment is applied to this mass in most cases, but it is not sufficient to purify raw sewage of organic matter which settles at the bottom, decays, and uses up so much oxygen that marine life dies. Or it may produce a rapid growth of algae which, when they die and their dead bodies sink to the bottom, produce large-scale fish kills, again because of the withdrawal of oxygen. During heavy storms not even the primary or secondary treatment can be applied to all the sewage because the amount of storm water mixing with the contents of the sewer systems exceeds the capacity of the treatment plants and the raw sewage pours directly into the Sound, causing heavy contamination. In such cases, all the soot, grime, oil and other waste are swept directly from the streets into the Sound.

The only way to solve the latter problem would be to separate the storm sewers from the regular drainage system to prevent overloading during rainstorms or when heavy snow melts suddenly. Laws passed by Congress require sufficient treatment of wastes by 1983 throughout the country so that all the waters of the nation will be preserved clean enough for swimming and other recreation; but the degree of purity will still be different from place to place and from time to time, depending on the weather. For the drainage systems of the cities cannot be separated easily; the cost is prohibitive. But much of the water could be recycled for irrigation, toilets, washing cars, and aquaculture.

At the Woods Hole (Massachusetts) Oceanographic Institution, experiments are being carried out with tertiary or triple-treated sewer water mixed with ocean water, in which oysters, sand worms and flounder are being raised. The method of raising oysters follows the Japanese system where strings hanging from a single raft produce four tons of oyster meat a year. At first single-celled marine plants, or diatoms, are raised in the solution; they are fed to the oysters. At the

201

bottom of the basins, sandworms feed on the excrement of the oysters. After they are fully grown, the sandworms are fed to the flounder in another basin. The final discharge of water into the sea after all these operations is as clear as the sea water itself. The only problem is to insure that no disease-causing pollutants contaminate the oysters. Hepatitis is one contaminant that can survive the triple treatment of the sewage. Six ponds, each half the size of a tennis court, are used for the experiments, and a power plant could, in the case of commercial application of this process, heat the water to any desired degree, producing even greater results.

Recreation versus Population Growth

If pollution from sewage is controlled, many more people than the five million who inhabited the Sound area in 1970 will want to utilize the waterway in the future. By 1990 the population of the region may reach seven million. Outside the immediate area, in New York City and the metropolitan district, there may be an additional ten to twelve million people looking for recreation. If mass transportation facilities are developed, the Sound area will be exposed to greater and greater pressures to open its beaches and other recreational facilities to the general public of the entire New York metropolitan region.

Yet of the total shoreline of the Sound, 413 miles, only 28 are still undeveloped, and most of that consists of narrow strips of beach below a bluff. At the same time the increased number of boat slips and moorings is an indication of the great demand for facilities besides public beaches. For one should realize that the 1,400 square miles of water surface of the Sound are well protected from storms, not like the ocean where pleasure craft can venture out only in good weather. Besides, the Sound offers excellent opportunities, at least potentially, for fishing, and its shallow waters were historically famous for producing the best crop of shellfish of all kinds. Thus, the demand for access to the waterway will grow, and it is evident that in the immediate future more areas of the Sound shore will have to be opened to the public. This will require the utilization of much land which today is used for other purposes, a difficult task in view of the fact that 85 percent of the shoreline of the Sound is privately owned and controlled. No wonder then that there is intense regulation of the beaches by the local towns with stickers required on cars and identification cards issued only to residents. Many of these towns refuse to accept federal aid for beach protection to combat erosion because it would imply an obligation for open admission and that would mean inundation by masses of people.

As we have seen in the section dealing with oil and the danger of oil pollution, commercial and industrial establishments now located on the waterfront might be required to move inland if the shoreline were to be freed for recreational purposes. With improved water quality resulting from pollution control, more recreational areas could be opened to the public. The transfer of privately owned and controlled land to public ownership would initially require large investments, but then, if well developed, public parks and facilities could be self-supporting with the public gladly paying the fees and tolls charged for that purpose. It would also be possible to establish some recreational facilities away from the Sound by creating man-made lakes inland. And finally, as we have seen above, the power companies could open up their cooling lakes, or parts of them, to the public. All in all, it is evident that only a large-scale, well-planned, well-administered recreation program can cope with the enormous increase in demand that the area will face in the not too distant future.

And large-scale extension of recreation facilities entails many related problems. Waterfronts may be cleared of industrial and port facilities, but the acquisition of large areas for parks is another question. The battles fought by Robert Moses in building up the park and recreational system of Long Island bear eloquent witness to this. Public parks fulfill their purpose only if there are adequate transportation facilities making them accessible to the masses. That means roads, parkways, and expressways and the resistance to them is growing everywhere, resistance not only by the wealthy but by whole communities which dread the prospect of heavy traffic with all its noise, crowds, and pollution. One way of lessening the traffic problem would be to provide bus lanes on existing roadways.

But even if every provision were made for lessening the effects of motor traffic and for building more parks and beaches, the remaining open areas around the Sound should be protected to save them from land speculators, either by direct acquisition or by other measures before they are subdivided. Moreover, zoning should be thoroughly studied so that the interest of the public can be served. Fortunately, a powerful movement is afoot to curb haphazard growth.

Suffolk County is planning now to preserve its farm areas from suburban sprawl. The county legislature is considering the purchase of 30,000 acres of farmland during the next decade at a cost of $150,000, or $5,000 per acre. The land would be rented to the farmers who would continue cultivating it, producing potatoes, cauliflower, and other high-quality crops. In this way, the farmers would not be tempted to sell their land to real estate promoters who would otherwise drive up the value of the acreage to the point where it would no longer be profitable to use it for farming. From 1965 to 1970 land values jumped from $1,867 per acre to $3,512 per acre, and the increase is continuing. Under this plan the remaining open spaces could be preserved for future generations and turned into recreational areas whenever the need arose, or whenever the money would be available. Time is getting shorter and shorter for such a measure because the dwindling of farmland in the county has assumed alarming proportions. In

203

1950 there were 123,000 agricultural acres in the county. In 1972 68,000 remained, of which only 10,000 were in the western part. At this rate there would be no farms left in Suffolk County in fifteen to twenty years simply because land for purposes of taxation is assessed without regard to use, the only criterion being the price land would bring on the open market. With soaring land values, the farmer is simply taxed out of farming and has to sell or go into real estate development of his own.

Most of the areas around the Sound face the same predicament as Suffolk County. A typical case is Guilford, Connecticut, one of the oldest colonial settlements. Until a few years ago the quiet atmosphere of a country town could be preserved there, but in 1970 work began on the first apartment building and the census of that year indicated that the population had more than doubled in the last decade, reaching 12,000. Parking places were becoming scarce around the village green, and the harbor began to fill up with pleasure boats. To preserve the woods and fields surrounding the town, the planning and zoning commission adopted a plan to zone home buildings in tight clusters. In 1965 a land conservation trust was formed to preserve open land by competing with developers in acquiring good land ordinarily swallowed up by urbanization. But how long can a community like Guilford resist the trend of the times? Located midway between New York and Boston and now practically a suburb of New Haven, it is threatened with submergence in megalopolis.

An even larger town, Westport, Connecticut, with thirty thousand inhabitants, decided to block the building of apartments in its territory, although an initial plan to permit the building of 340 luxury apartments had been accepted by the planning and zoning commission. This measure was reversed by the representative town meeting held on July 17, 1973. The reason was that this initial concession, which included sixty low-income apartments, would have marked the beginning of the end of the town's character as a quiet place of single-family suburban-type homes. The meeting of the representatives brought out the controversial nature of the issue. Some were for change, others for conservation. The latter carried the day. In the debate it was pointed out that the onion farmers, the original inhabitants of the town, opposed the coming of the artists, who in turn did not want the wealthy, and now the wealthy refuse to accept the coming of the junior executives, and all together are quite concerned about middle and low-income housing for people who work in or near the town. Bridgeport is only ten miles away to the east and Norwalk not far to the west. The pressure, as in the case of Guilford, is increasing, and the problem, temporarily banished, will certainly come back.

But it is not only fear of low-income masses threatening the semirural character of the wealthy Sound shore communities. Growth itself, even if it involves increasing the number of wealthy and middle class people in a town, has become taboo. A recent example of this zero growth attitude, previously discussed in chapter 9, was the refusal of Greenwich, Connecticut, to allow the giant Xerox Corporation to build its new headquarters in the town's territory even though the place selected for the building site was close to the Westchester County airport and situated far from the center of town. The reason offered by the town's planning and zoning commission was that permission would lead to the proliferation of office buildings and corporate headquarters in the country-side. Moreover, heavier use of the airport resulting from the influx of corporations would greatly increase noise pollution, and the large clerical staff working in the offices would increase the automobile traffic on streets already crowded during business hours.

Land Use and Development

Unless proper planning and development replace the haphazard spread of corporate headquarters, the Sound shore communities will face serious problems. The rapid increase of population dispersed all over the place in individual homes, and of campus-style office buildings, requires a vast network of connecting highways. This type of growth, if unchecked, will absorb the remaining open spaces too rapidly, and the existing parks and beaches will prove to be inadequate to accommodate all the people. Eventually the growing pressures will bring about a condition where only the local residents will be able to use recreation areas. This will certainly create tensions and controversies.

To avoid haphazard and chaotic growth the Nassau-Suffolk Regional Planning Board suggested a systematic channeling of development to prevent the disappearance of the still-existing open areas on Long Island. For housing the plan advocated the cluster principle that would bring apartments as well as individual homes closer together. Such an arrangement leaves much of the land in its natural condition, making it available for parks, playgrounds, ponds, and greenways, enhancing the quality of the environment. Such clusters would save pavement and relieve the cost of public services like fire and police protection and utilities. Much of the North Shore of the island could still be saved by such controlled development.

The principle of clustering or concentration could be applied to industry and commerce in the same manner by combining it with the corridor principle. This means that the so-called "spine" of the island, on which the Long Island Expressway and the main line of the Long Island Rail Roard run, could form a corridor of rapid communication and transportation around which centers of activity of all sorts could be established. Each center would be connected by bus lines with the area to the north and south of it. The Long Island Rail Road

205

could be improved to run high-speed trains between the centers and New York City.

Downtown districts in the old centers could be revitalized if the problem of cheap and swift mass transportation could be solved. As for travel by cars, increased parking accommodations could greatly help such urban renewal projects, for decay of downtown areas is due mainly to lack of such accommodations in the traffic-clogged streets and crowded parking lots. Brand-new supermarkets and shopping centers built far out in the country lured the people to their spacious parking lots, and the old business districts could not compete with them. Real estate developers built acres of suburban homes, subdividing the old potato and cabbage farms. As the middle class moved out of the towns, lower income groups moved in, especially into the districts adjoining the downtown sections, turning them into slums. To permit this situation to continue would be a waste of old values in the hearts of the towns. Restoration would save the central districts which formerly represented the focus of all activity. By creating better access to these areas, new housing with increased densities would stimulate the return of business and promote economic and social integration. Where older centers do not exist, new ones could be established with careful planning.

In this way a central corridor could be built through the counties of Nassau and Suffolk. Along its line would be centers of various sizes offering employment in their offices and factories, recreation in their cultural establishments, and shops and stores of the greatest variety. They would be situated mainly where the Long Island Rail Road crosses the main north-south highways. Smaller commercial and office centers could be located on the secondary corridors running through the northern and southern halves of the island, but with as little industry as possible. Wherever possible, mass transportation would be encouraged, with great density of housing centering near the railroad stations. More people could live closer to their jobs, and those who would still insist on the "one house, one lot" principle would have to pay a higher price.

Unless this overall picture is kept in mind and is coupled with careful planning and close supervision of growth, the area around the Sound will be doomed to overcrowding and the resulting social tensions. Its value as one of the great natural playgrounds of the nation and as a transportation route connecting some of the largest population centers will be lost or badly hurt, perhaps beyond repair.

An Exercise in Futuristics

Anyone making even a cursory analysis of the problems discussed in this chapter must surely ask himself wither goeth Long Island Sound? Will America's first urban sea be restored to its pristine beauty through the advances of modern technology, or will urbanization, industrialization, and suburban sprawl continue to encroach upon it? In attempting to answer this complex question one is perhaps wise to heed the incisive observations made by Robert Moses during a speech at Southampton in June 1973.[2]

Prosperity, education, new-found leisure and ambition to be with the "in" people have bred overnight experts on suburbia and exurbia. We produce them faster than hamsters. Our newly spawned intelligentsia are almost as prolific and fertile as the Chinese coolies who rise from the rice paddies to become atomic scientists. They also have bigger vocabularies.

Correct answers to the difficult question about Long Island Sound's future are hard to find because there are so many imponderables. The energy crisis which made its first big impact upon the American people in 1973, for example, may slow the antipollution drive. Power plants along the Sound heretofore opposed by the ecologists may be permitted. On the other hand, conservation of energy resulting in reduced auto usage may lead to improved air quality and may lessen the need for new Sound crossings. The introduction of large-scale mass transit facilities centering in the cities and towns of the Sound area may also reduce suburban sprawl, thereby preserving existing open spaces along the Sound for recreational use.

Implementation of a full-scale plan to rescue the Sound from man's carelessness will depend upon the willingness of the voters to fund the kind of clean-up programs recommended by the Long Island Sound Regional Study. Its preliminary report was issued in March 1974, and additional proposals may be included in the final report scheduled to appear in January 1975. In the final analysis, the price tag may simply be too big. On the other hand, partial adoption of the study's recommendations may be enough to turn the tide and guarantee Long Island Sound's survival for the future. Talk to a fisherman and he will tell you that the Sound *is* getting better. Improved water quality in the rivers flowing into the Sound because of the curtailment of factory waste disposal has resulted in the return of menhaden, or mossbunkers, to the estuaries of those rivers to feed on the plankton which now thrive there. Angry clashes between sport fishermen, seeking edible species, some of which feed on the menhaden, and commercial bunker boats out in search of the menhaden itself, made headlines in Sound shore newspapers during the summer of 1973. Early in 1974 the sport fishermen sponsored meetings in an effort to push for legislation limiting the activities of the bunker boats on the Sound.

The controversy between sport fishermen and commercial fishermen notwithstanding, the tempo of aquatic life in the Sound is picking up. This is a hopeful sign which, together with the increasing recognition on the part of Sound area residents that there is a limit to the growth an area can endure, augurs well for the future. Long Island Sound will never again be the unspoiled waterway traversed by Adriaen Block, but before the end of the twentieth century it can be a model urban sea free from serious pollution and open to millions of people who, hopefully, will use but not abuse our American Mediterranean.

Fishing for snappers from the rock jetty off Old Field Lighthouse (Photo by Allan M. Eddy Jr.)

208

Notes and References

CHAPTER 1

1. Daniel Denton, *A Brief Description of New York Formerly Called New Netherlands With the Places Thereto Adjoining,* ed. Gabriel Furman (New York: William Gowans, 1845), p. 2.
2. Adriaen Van der Donck, *A Description of New Netherlands,* ed. Thomas F. O'Donnell (Syracuse: Syracuse University Press, 1968), p. 55.
3. E. B. O'Callaghan, ed., *The Documentary History of the State of New York,* IV (Albany: Weed, Parsons & Co., 1851): 3.
4. Silas Wood, *A Sketch of the First Settlement of the Several Towns on Long Island* (Brooklyn: Alden Spooner, 1824), p. 35.
5. When Robert H. Roy, assistant United States district attorney, was searching the title of ownership of the island, the east end of which the government purchased for fortification purposes, he found the document among Southhold town records at Riverhead. The Indian chief had set his seal to it on April 27 1659 on Gardiners Island. The district attorney made a brief record without date of this interesting document entitled: "Plum Island History," of which a copy is preserved at Oysterponds Historical Society in Orient, L.I.
6. O'Callaghan, IV: 22.

CHAPTER 2

1. Henry P. Johnston, ed., *Memoirs of the Long Island Historical Society,* 3, *Campaign of 1776* (Brooklyn: Long Island Historical Society, 1878): 79.
2. "Nathan Hale" (pamphlet) (Huntington: Huntington Historical Society, 1933), pp. 18–19.
3. Peter Force, ed., *American Archives,* fifth series, 2 (Washington, D.C.: M. St. Clair Clarke & Peter Force, 1851): 991.
4. *Ibid.*

5. Otto Hufeland, *Westchester County during the American Revolution* (White Plains: Westchester County Historical Society, 1926), p. 105.

6. Charles Burr Todd, *In Olde Connecticut* (New York: The Grofton Press, 1906), p. 30.

7. William Wheeler, *The Journal of William Wheeler* in Cornelia P. Lathrop, ed., *Black Rock Seaport of Old Fairfield, Connecticut* (New Haven: Tuttle, Morehouse & Taylor Co., 1930), p. 29.

8. William Heath, *Heath's Memoirs of the American War* (New York: A. Wessels Company, 1904), p. 322.

9. John Warner Barber, ed., *Connecticut Historical Collections* (New Haven: Durrie & Peck and J. W. Barber, 1838), p. 276.

CHAPTER 3

1. *New York Evening Post,* March 25 1815, p. 1.

2. *Daily Herald* (New Haven), October 10 1833, p. 2.

3. Alfred Grant Walton, *Stamford Historical Sketches* (Stamford: Cunningham Press, 1922), p. 86.

4. *Long-Island Democrat* (Jamaica), January 22 1840, p. 3.

5. Nathaniel S. Prime, *A History of Long Island from Its First Settlement by Europeans to the Year 1845, with Especial References to Its Ecclesiastical Concerns* (New York: Robert Carter, 1845), p. 57.

6. *Ibid.*

7. *Daily Herald* (New Haven), July 29 1844, p. 2.

8. *Daily Herald,* August 5 1844, p. 2.

9. *Report of the Directors of the Long Island Railroad Company to the Stockholders* (Brooklyn: Long Island Rail Road Co., 1845), p. 9.

10. *Ibid.*

11. *Long-Island Democrat* (Jamaica), January 29 1840, p. 4.

CHAPTER 4

1. *New Haven Register,* December 29 1848, p. 2.

2. *Daily Palladium,* (New Haven), December 29 1848, p. 2.

3. *New Haven Register,* May 7 1853, p. 2.

4. Article reprinted in *Daily Palladium* (New Haven), May 7 1853, p. 2.

5. *Ibid.*

6. *Ibid.*

7. Frederick A. Hubbard, *Other Days in Greenwich* (New York: J. F. Tapley Co., 1913), p. 267.

8. E. B. Hinsdale, *History of the Long Island Rail Road Company: 1834–1898* (New York: The Evening Post Job Printing House, 1898), p. 8.

9. Peter Ross, *The History of Long Island from its Earliest Settlement to the Present Time* (New York: The Lewis Publishing Co., 1902), p. 297.

10. Long Island Rail Road Company, *Long Island* (New York: Aldine Press, 1882), p. 40.

11. *Glen Cove Gazette,* August 9 1873, p. 3.

12. H.C.A., "Plum Island History Fifty Years ago," typescript in Oysterponds Historical Society, Orient, L.I., p. 2.
13. *Ibid.*
14. *The Long-Island Democrat* (Jamaica), July 6 1880, p. 2.
15. Hubbard, p. 214.
16. *Glen Cove Gazette,* July 26 1873, p. 3.
17. Hubbard, p. 274.
18. W. W. Evans, *Letter to the President of the New-York and New Haven Railway Company with Reference to Railway Construction and Accidents* (New York: George F. Nesbitt & Co., 1864), pp. 10–11.
19. *Ibid.,* p. 11.
20. J. F. Scharf, *History of Westchester County,* 1 (Philadelphia: L. E. Preston & Co., 1886): 822.
21. *Ibid.*
22. Scharf, 1: 822.
23. *New York Times,* September 1 1893, p. 4.
24. Joseph Bucklin Bishop, ed., *Theodore Roosevelt's Letters to His Children* (New York: Charles Scribner's Sons, 1919), p. 59.
25. *Glen Cove Gazette,* July 12 1873, p. 2.

CHAPTER 5

1. Ulysses Prentiss Hedrick, *A History of Agriculture in the State of New York* (New York: New York State Agricultural Society, 1933), p. 372.
2. Timothy Dwight, *Travels in New England and New York,* 3 (London: William Baynes & Son, 1823): 285.
3. *Ibid.,* 278.
4. Benjamin F. Thompson, *History of Long Island,* 3 (New York: E. French, 1918): 245.
5. Edith Loring Fullerton, *Peace and Plenty. The Lure of the Land. A Call to Long Island. The Story of the Work of the Long Island Railroad Company at Experimental Station Number One.* (Jamaica: Long Island Rail Road Company, 1906), p. 150.
6. Horatio Gates Spafford, *A Gazetteer of the State of New York* (Albany: B. D. Packard, 1824), p. 561.
7. Hedrick, p. 165.
8. Nathaniel S. Prime, *A History of Long Island from Its First Settlement by Europeans to the Year 1845 with Especial References to Its Ecclesiastical Concerns* (New York: Robert Carter, 1845), p. 75.
9. *Ibid.,* p. 74.
10. Ebenezer Emmons, *Agriculture of New York* (Albany: C. Van Benthuysen, 1851), p. 267.
11. *Ibid.*
12. Letter from Captain Frederic A. Weld to David G. Floyd, November 29 1845, Oysterponds Historical Society, Orient, L.I.
13. F. M. Caulkins, *History of New London, Connecticut, from the First Survey of the Coast in 1612 to 1860* (New London, H. D. Wiley, 1895), p. 644.
14. *Ibid.*
15. Dwight, p. 291.

16. Eugene G. Blackford, *Second Report of the Oyster Investigation of the Oyster Territory for the Years 1885 and 1886* (Albany: The Argus Co., 1887), p. 5.
17. *Ibid.,* p. 6.
18. *Ibid.,* p. 18.
19. Spencer P. Mead, *Ye Historie of ye Town of Greenwich* (New York: Knickerbocker Press, 1911), p. 338.
20. D. Hamilton Hurd, *History of Fairfield County* (Philadelphia: J. W. Lewis & Co., 1881), p. 100.
21. *Ibid.*

CHAPTER 6

1. *New York Times,* January 11 1914, 8:3.
2. *Ibid.*
3. *New York Times,* February 28 1915, 8:6.
4. *New York Times,* April 13 1913, 8:6.
5. *New York Times,* April 9 1916, 3:7.
6. *New York Times,* April 25 1913, p. 10.
7. "A Week-End at Kensington," (Kensington: Rickert-Finlay Realty Co., 1915), pamphlet in Kensington File, Nassau County Historical Museum Library, East Meadow, L.I.
8. *Ibid.*
9. *New York Times,* January 15 1906, p. 1.
10. *New York Times,* April 22 1917, p. 16.
11. *New York Times,* March 26 1918, 3:6.

CHAPTER 7

1. F. Scott Fitzgerald, *The Great Gatsby* (New York: Charles Scribner's Sons, 1953), p. 49.
2. *Ibid.,* p. 39.
3. *New York Times,* April 7 1924, p. 19.
4. *New York Times,* June 20 1926, 4:6.
5. Richard Schermerhorn, Jr., *Master Plan of the Great Neck District* (Great Neck: The Great Neck Association, 1927), p. 1.
6. *New York Times,* February 7 1926, 10:14.
7. *New York Times,* February 17 1924, 9:2.
8. *New York Times,* January 4 1924, 11:2.
9. *New York Times,* May 8 1921, 9:1.
10. *New York Times,* April 21 1925, p. 22.
11. *New York Times,* January 7 1926, p. 11.
12. *New York Times,* January 11 1926, p. 26.
13. The governor's annual message was reprinted in the *New York Times,* January 7 1926, p. 11.
14. Fitzgerald, p. 182.

CHAPTER 8

1. *New York Times,* January 3 1937, 2:1.

2. Howard W. Palmer, "Present and Future" in *Greenwich Old and New* (Greenwich: The Greenwich Press, 1935), p. 159.
3. *New York Times,* April 4 1937, p. 12.
4. *Ibid.*
5. Arthur L. Hodges, "America's Last Frontier," *Long Island Forum,* 2 (October 1939): 17ff.
6. *Newsday,* April 10 1945, p. 30.

CHAPTER 9

1. Robert Moses, *Vital Gaps in New York Metropolitan Arteries* (New York: Triborough Bridge and Tunnel Authority, 1940), p. 1.
2. New Haven Board of Aldermen, *Verbatim Proceedings: Public Hearing Aldermanic Committee on Streets and Squares: Long Wharf Redevelopment Plan,* October 16 1958.

CHAPTER 10

1. New England River Basins Commission, *Plan of Study: Long Island Sound Regional Study* (Boston: New England River Basins Commission, 1971), p. 1–1.
2. *Loc. cit.*
3. Subcommittee on Executive Reorganization and Government Research of the Committee on Government Operations, United States Senate, Ninety-First Congress, Second Session on S.2472, Hearings: *Preserving the Future of Long Island Sound:* July 1970, part 3 (Washington, D.C.: U.S. Government Printing Office, 1970, p. 382.
4. *New York Times,* June 14 1970, p. 70.
5. *New York Times,* August 3 1956, p. 21.
6. *Loc. cit.*
7. *Preserving the Future of Long Island Sound,* part 3, p. 376.
8. *Loc. cit.*
9. *Preserving the Future of Long Island Sound,* part 3, p. 467.
10. *Ibid.,* p. 464.
11. *Ibid.,* p. 394.
12. *Ibid.,* p. 396.

CHAPTER 11

1. *New York Times,* April 30 1939, p. 37.
2. *New York Times,* April 30 1939, p. 1.
3. *New York Times,* April 30 1939, p. 37.
4. *Ibid.*
5. *New York Times,* November 2 1937, p. 27.
6. *New York Times,* April 21 1955, p. 19.
7. *New York Times,* October 23 1957, p. 21.
8. *New York Times,* March 6 1965, p. 24.
9. *Traffic, Earnings and Feasibility of the Long Island Sound Crossing,* August 2 1965, p. 1 (booklet containing Robert Moses' comments on the Madigan-Hyland report).
10. *Ibid.,* p. 2.

11. Bertram D. Tallamy Associates, *Feasibility Report, Highway-Railroad Crossings Suffolk County, New York, to Connecticut,* February 1968, ii.
12. *New York Times,* March 23 1967, p. 24.
13. *New York Times,* February 4 1969, p. 54.
14. Creighton, Hamburg, Inc., *A Comprehensive Transportation Study for Proposed Bridge Crossings,* December 1971, p. 5.
15. *New York Times,* January 31 1972, p. 82.
16. R. Q. Praeger, letter transmitting to the N.Y. State Dept. of Transportation, Traffic, Earnings and Feasibility of the Long Island Sound Crossing between Rye and Oyster Bay, December 1968.
17. New York State Department of Transportation, *Long Island Sound Crossings and Approach Highways,* November 1972, p. 59.
18. Letter from Raymond T. Schuler, Commissioner of Transportation, New York State, April 16 1973.
19. Metropolitan Transportation Authority and New York State Department of Transportation, *Long Island Sound Crossing* (booklet), December 1972, p. 21.
20. *Ibid.*
21. *New York Times,* December 5 1965, p. 68.
22. *New York Times,* March 28 1967, p. 44.
23. *New York Times,* February 11 1968, 13:2.
24. *Ibid.*
25. Press release from Citizens for Sound Planning, September 1972.
26. *Ibid.*
27. *Ibid.*
28. *New York Times,* July 18 1965, p. 58.

CHAPTER 12

1. Long Island Sound Regional Study, *Electric Power Generation:* An Interim Report (New Haven, NERBC, 1973), p. 38.
2. Robert Moses, "The Future of Eastern Long Island," June 21 1973, p. 3.

Bibliographic Note

Published material on the history of Long Island Sound is scattered in a wide variety of books, periodicals, and newspapers. Of the books dealing specifically with the Sound, a limited amount of historical data can be found in F. S. Blanchard, *Long Island Sound* (1958), and Julius M. Wilensky, *Where to Go, What to Do, How to Do It on Long Island Sound* (1971). Morton Hunt's well written account of a voyage around the Sound in the early sixties, *The Inland Sea* (1965), contains some interesting bits of history. Additional historical information appears in Charles Hervey Townshend, *The Early History of Long Island Sound and Its Approaches* (1894) and in *The Commercial Interests of Long Island Sound and New Haven in Particular* (1883) by the same author.

An historical account of at least one aspect of the Sound's past, the era of the steamboats, is Roger McAdam's *Salts of the Sound* (1939, 1957). McAdam's other books, *Commonwealth: Giantess of the Sound* (1959), and *Priscilla of Fall River* (1956), are also helpful, as are Samuel Ward Stanton, *Long Island and Narragansett Bay Steam Vessels: Drawings* (1962), and Henry Whittemore, *The Past and Present of Steam Navigation on Long Island Sound* (1892). Shipwrecks in the Long Island Sound area are described in Jeanette Edwards Rattray, *Ship Ashore* (1955).

For the reader interested in published accounts of the history of the various Sound shore communities, a good starting place would be the standard histories of the states and counties bordering the Sound. The following histories of Connecticut provide information about the Sound areas of the state: George L. Clark, *A History of Connecticut* (1914); G. B. Hollister, *The History of Connecticut;* Norris G. Osborn, *History of Connecticut* (1925); Benjamin Trumbull, *A Complete History of Connecticut* (1797), Albert E. Van Dusen, *Connecticut* (1961). For material about the Sound coast of Westchester, the standard histories of the county yield considerable information. These include: Robert Bolton, *History of the County of Westchester* (1848); Alvah P. French, *History of Westchester County, New York* (1925–27); Thomas J. Scharf, *History of Westchester County* (1886); F. Shonnard and W. W. Spooner, *History of Westchester County* (1900).

Moving across the Sound to Long Island, we find no dearth of historical information. The older histories such as Benjamin F. Thompson, *History of Long Island,* 3 vols. (1839, 1918), and Nathaniel S. Prime, *A History of Long Island from Its First Settlement by Europeans to the Year 1845 with Especial References to Its Ecclesiastical Concerns* (1845) are still valuable, provided that the reader keeps in mind that they were written well over one hundred years ago. Although the Thompson work was updated in 1918, it remains a product, albeit a good one, of its time. More recent histories of the island containing information about the Sound are: Ralph H. Gabriel, *The Evolution of Long Island* (1921, reprinted by Kennikat Press 1960), an excellent, almost timeless work, and Paul Bailey, ed., *Long Island: A History of Two Great Counties: Nassau and Suffolk,* 3 vols. (1949).

In addition to these general works the serious scholar will wish to consult the following newspapers:

LONG ISLAND	MAINLAND
Brooklyn Eagle	*Daily Herald* (New Haven)
Glen Cove Gazette	*Daily Item* (Port Chester)
Hempstead Inquirer	*Daily Palladium* (New Haven)
Long Island Democrat (Jamaica)	*Daily Times* (Mamaroneck)
Long Island Press (Jamaica)	*Hartford Times*
Newsday (Garden City)	*New Haven Register*
Oyster Bay Guardian	*New York Times*
	Rye Chronicle
	Stamford Advocate
	Standard Star (New Rochelle)
	Stratford Times

Most of the Long Island papers are available on microfilm at the library of the Nassau County Museum, Eisenhower Park, East Meadow, Long Island. The Long Island Historical Society, Brooklyn, New York, and the Queensborough Public Library's Long Island Division in Jamaica also have extensive collections of Island papers. Newspapers published on the mainland side of the Sound can be found at the New Haven Colony Historical Society, the New London County Historical Society, the Fairfield County Historical Society, the Westchester County Historical Society, Tuckahoe, New York, as well as at public libraries in the Sound shore communities.

Certain periodicals also provide information about various aspects of the

Sound's development and particularly about the ecological problems plaguing the waterway in modern times. For this aspect the interested reader is advised to consult the following magazines: *The Conservationist, On the Sound, National Geographic, Scientific American.* The following magazines are also useful for gaining insight into the Sound and the Sound shore communities: *Long Island Forum, Nassau County Historical Society Journal, Westchester Historian, Yachting, Travel, Country Life, Environment.*

Helpful though they are, books, newspapers, and periodicals do not tell the full story. For this, one has to ferret out the treasures buried in various archives around the Sound. Several of the aforementioned county historical societies have not only an array of excellent published material for their areas but useful unpublished documents as well. The societies and their manuscripts and other specialized material relating to the Sound and the Sound shore communities are:

MAINLAND

New Haven Colony Historical Society: Manuscripts on the following subjects relating to the Sound and the Sound shore communities: Long Wharf, Steamboat *Robert Fulton,* Oystering, the Farmington Canal, the Amistad Case, miscellaneous shipping and customs records, Revolutionary War pension records, deeds for Connecticut and those parts of Long Island included in the Connecticut Colony.

Fairfield County Historical Society: Extensive card files on ships sailing out of Fairfield County ports, scrapbooks containing illustrations of steam yachts, clipping files on towns in Fairfield County, mainly from Fairfield north, journal of Eleazar Bulkley containing firsthand accounts of Revolutionary War events, journal of Isaac Jennings detailing voyages from the Sound to Europe, South America, and the West Indies.

New London County Historical Society: Manuscript material on New London's role as a seafaring community. Of interest are the log books of the following vessels: *Wanderer, Florence, Venice, Columbia;* and the Billings papers consisting of letters, accounts of condition of ships, orders to various masters from the firm.

Westchester County Historical Society: Clipping files for the Sound shore communities, standard histories of Westchester and of the towns and villages along the Sound.

LONG ISLAND

Library of the Nassau County Museum, formerly library of the Nassau County Historical Society: In addition to a sizable and continually updated collection of secondary sources, the library has an excellent clipping file for every community in the county. The following materials relating to the Sound and Sound shore communities are also included in the library's collection: Oyster Bay land surveys and deeds for the eighteenth century; Long Island auto maps c. 1910; World War I booklets; annual reports of the Queens Nassau Agricultural Society for selected years between 1870 and 1925.

Suffolk County Historical Society: Log books of coastal voyages in the nineteenth century, files on lighthouses, ferries and steamers, whaling, as well as a fine collection of the principal secondary sources on Long Island history.

Other specialized collections on the island include:

Morton Pennypacker Collection, East Hampton Library: Secondary sources on Long Island history plus the works of Charles Hervey Townshend on the history of the Sound and selected periodicals containing articles on the Sound.

The Richard Handley Collection, Smithtown Library: Published material on the history of Long Island but little manuscript or other documentary material pertaining to the Sound.

Grist Mill Historical Collection, William Cullen Bryant Library, Roslyn: Fascinating scrapbooks and other memorabilia of the famous Mackay family who lived in Roslyn during the height of the Gold Coast era; letters and other mementos of William C. Bryant.

The Huntington Public Library: Good collection of secondary sources plus unpublished student papers on the Revolutionary War in Huntington and other local topics.

Huntington Historical Society: Manuscripts relating to whaling and shipping; a number of bills of sale for schooners, plus secondary sources on Long Island history and a good collection of photographs.

Town of Brookhaven Historical Collection, Patchogue, books and files for the towns along the north shore of Brookhaven; nineteenth-century atlases and maps.

Cold Spring Harbor Whaling Museum: manuscripts on Sound shipping and the whaling industry of Cold Spring Harbor.

Oyster Bay Historical Society: manuscripts relating to shipping; freight records of the Northport, Huntington, and Oyster Bay Steamboat Company.

Oysterponds Historical Society, Orient: an excellent collection of primary and secondary sources classified according to names and kinds of documents but not according to subject. Materials consulted included: accounts of the 1938 hurricane; diary of an unknown farmer (1864–65); Augustus Griffin's journal; documents pertaining to shipping and Sag Harbor customs house; letters discussing Orient and the Sound during the War of 1812.

Queensborough Public Library, Long Island Division: An excellent collection of the principal secondary sources on Long Island history, plus Long Island newspapers and clipping files on pertinent topics related to the island's past.

Local historical societies on the mainland yielding information about the Sound were:

The Hufeland Library, Huguenot and Thomas Paine Historical Association, New Rochelle: Although not fully catalogued, this excellent small library, once the private collection of Westchester historian Otto Hufeland, contains almost every publication on Westchester County through the 1920s as well as New Rochelle newspapers, scrapbooks for Westchester communities, and an extensive collection of books and maps pertaining to the Revolutionary War.

Rye Historical Society: The Slater papers, manuscript material pertaining to Rye during the eighteenth and early nineteenth centuries; legal documents, business papers, shipping records, including bills of sale for the original Rye–Oyster Bay ferry.

Mainland libraries whose collections were investigated included:

The Pequot Library, Southport, Connecticut: An extensive collection of secondary sources, plus unpublished student papers on Connecticut history.

Phoebe Noyes Griffin Library, East Lyme, Connecticut: A good collection of secondary sources, including histories of the local region.

Beinecke Rare Book Library, Yale University: A number of interesting items pertaining to the Sound and the Sound shore communities as well as a collection of eighteenth- and nineteenth-century newspapers whose originals scholars may consult, thereby eliminating the need to use eye-straining microfilm machines. In addition to newspapers the following sources were consulted: pamphlet material on the Farmington Canal; Long Island Railroad Stockholders Report (1837); Long Island Railroad promotional pamphlets.

Although rather far removed from the Sound, the *Connecticut State Library* at Hartford has a variety of material relating to the waterway. Among the sources consulted there were: miscellaneous papers on the New York, New Haven and

Hartford Railroad; the Henry White papers, which contain some material on the New Haven Railroad; logs of coastal voyages and whaling trips.

The following libraries and historical societies located beyond the Sound area also have information about the inland sea:

Rhode Island Historical Society, Providence: Pamphlet material on Long Island Sound steam navigation.

Brown University Library, Providence: Letter of Rochambeau concerning plans to clear Long Island Sound of enemy refugees in 1781; document c. 1674 giving Connecticut commercial and political privileges on eastern Long Island.

The collections of the *New York Public Library* at 42nd Street and Fifth Avenue, New York City, the *New York Historical Society* on Central Park West, New York City, and the *Long Island Historical Society* in Brooklyn contain a wealth of material pertaining to the history of Long Island.

Last but not least among the historical societies and libraries, the library of the *Marine Historical Association,* Mystic, Connecticut, has a considerable amount of material dealing with the Sound. Among the sources consulted there were: records of the New England Navigation Company, the Norwich, New London and Block Island Steamboat Company, the Old Colony Steamboat Company, Nathaniel M. Allen papers containing logs of ships operating between Providence and New York, journals describing whaling voyages on the following vessels: *France, Washington, Concordia, Prudent, Triad,* and *Catherine Thomas.*

A number of historical societies in the Sound area have materials of local interest including newspapers and local history files but at present do not have extensive collections of Sound-related data unavailable elsewhere. For the serious researcher in quest of primary source material, the following libraries and societies are of limited value:

John Jermain Memorial Library, Sag Harbor, L.I.
Southold Free Library, Southold, L.I.
Bridgeport Public Library, Bridgeport, Connecticut
Stamford Historical Society, Stamford, Connecticut

In a less historical vein material about the current problems of the Sound was obtained from the following governmental agencies:

Atomic Energy Commission, Washington, D.C.
U.S. Department of Agriculture, Soil Conservation Service, Riverhead, L.I.
U.S. Department of Commerce, National Oceanic and Atmospheric Adminis-

tration, National Marine Fisheries Service, Water Resources Division, Gloucester, Massachusetts

U.S. Department of the Interior, Geological Survey, Water Resources Division, Hartford, Connecticut

U.S. Department of the Interior, Fish and Wildlife Service, Patchogue, L.I.

U.S. Department of Housing and Urban Development, Boston, Massachusetts

State of Connecticut, Department of Environmental Protection, Hartford, Connecticut

State of Connecticut, Office of State Planning, Hartford, Connecticut

City of Stamford, Urban Redevelopment Commission, Stamford, Connecticut

Bridgeport Redevelopment Agency, Bridgeport, Connecticut

State of New York, Public Service Commission, Albany, New York

New York State Department of Environmental Conservation, Albany, New York

Department of Parks, Recreation and Conservation, Westchester County, White Plains, New York

New England River Basins Commission, Long Island Sound Study, New Haven, Connecticut

1790 House in Cold Spring Harbor (Photo by the author)

Index

INDEX

Tugboats at Port Jefferson Harbor (Photo by Allan M. Eddy Jr.)

228